RESTLESS EARTH

Prepared by
The Book Division
National Geographic Society
Washington, D.C.

RESTLESS EARTH

Contributing Authors
H. J. de Blij
Michael H. Glantz
Stephen L. Harris
Patrick Hughes
Richard Lipkin
Jeff Rosenfeld
Richard S. Williams, Jr.

Published by
The National Geographic Society
Reg Murphy, *President*
  and Chief Executive Officer
Gilbert M. Grosvenor, *Chairman of the Board*
Nina D. Hoffman, *Senior Vice President*

Prepared by The Book Division
William R. Gray, *Vice President and Director*
Charles Kogod, *Assistant Director*
Barbara A. Payne, *Editorial Director*
  and Managing Editor

Staff for this book
Bonnie S. Lawrence, *Project Editor*
  and Senior Researcher
Carolinda E. Hill, *Text Editor*
John G. Agnone, *Illustrations Editor*
Jody Bolt Littlehales, *Art Director*
Victoria Garrett Jones, Kimberly A. Kostyal,
  Researchers
Joseph F. Ochlak, *Map Researcher*
Martin S. Walz, *Map Production*
Carolinda E. Hill, Bonnie S. Lawrence,
  Peyton H. Moss, Jr., *Contributing Writers*

Lewis R. Bassford, *Production Project Manager*
Richard S. Wain, *Production*

Meredith C. Wilcox, *Illustrations Assistant*
Kevin G. Craig, Dale-Marie Herring,
  Peggy J. Purdy, *Staff Assistants*

Manufacturing and Quality Management
George V. White, *Director*
John T. Dunn, *Associate Director*
Vincent P. Ryan, *Manager*
Polly P. Tompkins, *Executive Assistant*

Anne Marie Houppert, *Indexer*

Library of Congress CIP data: page 284

*OPPOSITE: Clouds of ash billow from a volcano in
Guatemala. Explosive volcanic eruptions release ash
and gases that can affect world climate.*

*PAGE ONE: Rain cascades over the coastal forest near
Ketchikan, Alaska. At any given moment some 2,000
thunderstorms take place on the planet.*

*PRECEDING PAGES: A computer-enhanced satellite image
shows powerful Hurricane Fran approaching the East
Coast of the United States in September 1996.*

*FOLLOWING PAGES: Huge waves hit the California coast
in January 1988 (upper left). Wind causes most waves,
but earthquakes and volcanic eruptions trigger others. A
mass of clouds advances across fields near Miami, Texas
(upper right). Such storms can spawn tornadoes. In 1995,
a moment-magnitude 6.9 quake struck Kobe, Japan
(lower left). Some modern buildings sagged and showered
the city's streets with glass. A woman in Mauritania
waters her garden (lower right). Drought devastated
Africa's Sahel region in the 1970s and '80s.*

CONTENTS

DAVID OLSEN/WEATHERSTOCK

ASAHI SHIMBUN

WARREN FAIDLEY/WEATHERSTOCK

STEVE McCURRY

FOREWORD

When America's astronauts stood on the moon for the first time and looked back across the void, they watched the earth, serene and silent, rising above the lifeless lunar surface. Great swirls of white clouds veiled the familiar shapes of brownish and greenish continents set in blue oceans, with the Antarctic ice sheet the brightest of all. Greenish patches on the continents drew their eyes toward these oases of life on the earth.

But the astronauts knew that their planet is anything but serene or silent. The familiar continental outlines are not immutable but have changed repeatedly during the nearly five billion years of the earth's existence. Great mountain ranges have been created and worn down; coasts have been established and eroded.

Earthquakes rumble through the crust and upper mantle, as rock formations are fractured by tectonic processes. Volcanoes spew gases, tephra, and aerosols into the atmosphere while they inundate the land and seafloor with lava flows and build hills, mountains, and ridges. And all the time, gravity tugs at every rock fragment and water molecule, slowly modifying every landform. Sometimes even that deceptively quiet process has catastrophic outcomes. Mudslides, debris avalanches, landslides, and snow avalanches ravage mountainous regions, often causing death and destruction.

Who among us has never come face to face with nature's fury? Is there a city dweller or a suburbanite whose daily commute has not been defeated by a blizzard, severe thunderstorm, flash flood, or hurricane? Is there a farmer who has not had to cope with flood or drought, a seafarer who has not been threatened by a tempest? Wherever we live and work, we are all vulnerable to the stirrings of the restless earth. No place is immune to the effects of nature's power.

The earth is restless because of the dynamic nature of its crustal plates, mantle, and core. The planet continues to transfer heat from its interior to its surface, while slowly shifting the 16 major plates that make up its relatively thin crust. This activity is driven by forces far beneath the surface, where geochemical processes and high temperatures keep rock mobile.

At continental converging margins, crustal plates meet along thousands of miles. In stupendous collisions the plunging plates push up the overlying plates, building spectacular mountain ranges and setting the stage for the sculpturing of scenic coastlines. Great fractures along these collision belts mark the location of recurrent earthquakes; molten lava along such a crustal boundary often moves up through the overlying plate, creating spectacular volcanoes, such as Mount Rainier or Mount St. Helens in the Pacific Northwest.

Regions of volcanic activity and earthquakes are often closely related, as any map of the physical geography of the Pacific Ocean indicates. The Ring of Fire nearly encircling the Pacific is aptly named, because this zone contains not only the greatest concentration of earthquake activity but also the majority of the planet's more than 500 known active volcanoes on land. Although not on the same scale as earthquakes, where loss of life may exceed 655,000 (Tangshan, China, 1976) and property damage may run into billions of dollars (Kobe, Japan, 1995), volcanoes may also cause great property damage and exact a toll of human life. Volcanoes, however, often issue warning signs that eruptions are in the offing, and volcanologists have learned, for many of them, to decipher premonitory signs of impending eruptions. During the 20th century, as many as 50,000 people in the Western Hemisphere could have been saved if forewarnings from Mount Pelée on Martinique and Nevado del Ruiz in Colombia had been heeded by government officials.

The earth's oceans and seas cover more than 70 percent of its surface, and are kept in motion by a combination of processes originating in the atmosphere, the sun, and the planet's rotation. These forces propel great currents into circulation systems that move vast volumes of seawater clockwise in the Northern Hemisphere, counterclockwise in the Southern Hemisphere. This "conveyor" system transports warmer ocean waters from lower to higher latitudes, expanding the habitable world.

When a cyclonic storm such as a hurricane travels across the ocean, a dome of water rises under its central low-pressure area. Most property damage is caused by this surge of wind-whipped water as a storm makes landfall, but high winds and tornadoes also contribute to losses. With polar-orbiting and geostationary satellites and good communication networks, the loss of life from hurricanes in the Caribbean Sea, Gulf of Mexico, and eastern United States has been markedly reduced. Not so in poorer countries such as Bangladesh, where the death toll from a cyclone sweeping in from the Bay of Bengal can still reach hundreds of thousands.

We humans cluster on the surface of the earth, but we also live at the bottom of an ocean of air. The atmosphere blankets land and sea, deriving moisture from the oceans, lakes, and plant cover. Great maritime and continental air masses move across the land and water, and the clash between cold, dry air and warm, moist air often produces thunderstorms and tornadoes; unstable warm air also gives rise to storms. Daily weather and long-term climate variability are the results of a dynamic atmosphere, the principal source of most people's daily interaction with their natural environment.

The historical geography of human civilizations is the story of adaptation to environmental challenges, among which climate takes the lead. Today, technology allows people to live under virtually any natural condition if the society can afford the cost. For millions of city dwellers in wealthy nations, weather and climate have receded as factors in life and survival. But for billions of other people, they remain the critical component in subsistence and survival.

Restless Earth describes how floods overpower human engineering in one part of the world even as droughts take their toll in another. Long-term climate fluctuations plunge parts of the planet into periods of desertification, resulting in dislocation of humans and natural flora and fauna. As population grows, the risks increase, because humans are spreading into marginal lands or becoming ever more densely packed in regions prone to floods, droughts, severe storms, and other natural hazards.

The increasing incidence of devastating fires in the western United States reflects exactly this combination, exacerbated by the fire-suppression activities of well-meaning local,

state, and federal agencies that allow the dangerous accumulation of forest litter. Drought, which is a natural part of the prevailing climate regime in many regions, is certainly a factor; another is the penetration of woodlands by suburbanization and the planting of non-native species, such as eucalyptus trees. These drought-resistant trees, endemic to Australia, contain an oil that is extremely flammable.

By global standards, the U.S. population is not growing rapidly, but when it comes to the dispersal of people, the U.S. has no peer. Good roads, inexpensive gasoline, and high incomes combine to push the suburban "front" into landscapes that are often vulnerable to the forces of nature.

Some wildfires are set by arsonists, but many more result from nature's own fire starter: lightning. In many regions a wildfire is a natural process that rejuvenates the landscape and allows flora and fauna to flourish. In Australia, firestorms racing through the crowns

GARY BRAASCH/TONY STONE IMAGES

of eucalyptus trees frequently devastate regions. When such a fire came to Hunter Valley in 1978, visitors stood in awe while it swept across a ridge and descended as a glowing avalanche of flame and smoke into the lowland vineyards. The fire blocked the sun and filled the air with a noxious yellow haze that wilted vines even before they were burned. The terror of a fast-racing wildfire is similar to that of a tornado: Speed and sudden changes in direction trap victims before they have a chance to escape.

Restless Earth provides a powerful reminder of the planet's capricious forces and processes, of nature's challenges and humanity's varied and often futile responses. Overshadowing our daily routines is the latent possibility that an unexpected act of nature will change our lives forever. It has happened before, and it will happen again.

H. J. de Blij and Richard S. Williams, Jr.

OUR LIVING EARTH
ITS RAGING FORCES

by Richard Lipkin

INTRODUCTION

In the blackness of the universe, far from billions of more brilliant suns, an average star in a nondescript galaxy showers warm, energetic light onto nine planets circling about it in space. Out of the debris from once neighboring, but now exploded, stars the four gas giants and five small rocky orbs that make up our solar system came into being, including the one that we call the earth.

Ever since it coalesced more than 4.6 billion years ago, the earth has been constantly changing. A dynamic planet, it may suddenly experience a violent geologic upheaval—a great movement of the crust in an earthquake or an explosive volcanic eruption. And all the while, the slower movement of its crustal plates rearranges the continent-ocean geometry over tens of millions of years.

"Earth is living and dying, all the time creating and destroying its own surface," newscaster Walter Cronkite observed while reflecting on the earth's magnificence. "It is a body in motion, alive—and in its vitality, awesome.

"Compared to the great energy that Earth unleashes, humankind is insignificant," Cronkite also noted. "We haven't built a nuclear weapon with the explosive power of a Mount St. Helens; with all our chemicals and machines we are helpless midgets before the moving Earth."

About two-thirds of the way into the roughly 15-billion-year history of the universe, our solar system and all that we know of its present contents emerged from a dense cloud of hydrogen and dust—compressed, many scientists now believe, by the shock wave from a nearby exploding star, or supernova, which may have simultaneously seeded the solar system with heavier elements. From an enormous spinning disk of gas and particles, the sun and its planets slowly coalesced.

In time, the earth began to sort the semi-molten mass of accumulating planetesimals into rocks of different density. As orbiting debris smashed into the newly formed planet, gravitational forces and internal radioactive processes heated its interior to nearly 2000°F. Metals such as iron and nickel sank and formed a solid inner core surrounded by a molten outer core, while less-dense silicates formed the mantle and crust.

Newborn volcanoes emitted water vapor and other gases, as well as aerosols, lava, and tephra (fragments of solid material). From within the earth's mantle, convection currents in semi-molten rock brought magma into the crust; where it reached the surface, lava flowed out and slowly cooled. Released from the cooling lava, water vapor and other gases, including carbon dioxide and nitrogen, formed a proto-atmosphere because the gases were held close to the earth by the force of gravity. Water vapor condensed and rain began to fall, filling rivers and streams that eroded the land and then flowed into the proto-oceans. Finally, several hundred million years following coalescence from the solar nebula, the third planet from the sun had oceans, land, and an atmosphere.

About three and a half billion years ago, in this low-oxygen, ultraviolet-intense environment, the earliest life-forms—simple, unnucleated cells known as prokaryotes—emerged. Somehow, scientists believe, sparked by lightning and spurred by ultraviolet light from the sun, amino acids formed. They came together,

In 1972, Harrison H. Schmitt, a geologist and an astronaut, snapped this photograph from Apollo 17 while the space capsule orbited at a distance of 25,000 miles. Swirling clouds blanket part of snow- and ice-covered Antarctica, as well as parts of Africa and Asia, while sparkling blue oceans span the areas between the continents. Beyond the Red Sea, a great desert reaches across the Arabian Peninsula. At any moment clouds cover roughly half the earth's surface, reflecting some solar energy back into space while simultaneously trapping atmospheric heat.

PRECEDING PAGES: Lightning strikes near northern California's Lava Beds National Monument.

probably on rocks near tidal pools, and forged links. In tidal pools protected from harsh light, they produced complex molecules capable of self-replication.

While the earliest cells lived in the oceans and absorbed nutrients for food, later organisms learned to capture sunlight. Through a process called photosynthesis, these organisms were able to make their own food. Blue-green algae, which are known as cyanobacteria today, formed "stromatolite" colonies and gave off oxygen as a by-product. What followed forever changed the planet's course of evolutionary

biology: Oxygen-based life-forms developed and rapidly spread, first in the oceans and later to the land.

To sustain the remarkable abundance of organisms that eventually appeared in a variety of shapes, colors, and sizes, the earth had to undergo substantial changes. Green plants slowly spread across the land, markedly increasing the concentration of oxygen in the earth's atmosphere. Oxygen atoms linked together and over time formed a protective ozone layer in the upper atmosphere, thus shielding the earth's surface and its developing

life-forms from the harmful effects of the sun's ultraviolet rays.

Surrounded by an atmosphere rich in oxygen and oceans rich in nutrients, life-forms evolved ways to use oxygen to get energy from food, opening a pathway for new varieties of organisms to flourish, such as fish, amphibians, reptiles, birds, and mammals. The so-called geosphere—including the atmosphere (gaseous envelope), lithosphere (solid, rocky crust and upper mantle), hydrosphere (water in rivers, lakes, and oceans), and cryosphere (water in the form of sea ice, glaciers, snow, and frozen ground)—forms the natural environment for the planet's five great kingdoms of life to flourish (the biosphere).

From the densest part of the core to the thinnest part of the upper atmosphere, the earth is made up of layers. The low-lying troposphere, ranging in thickness from 12 miles in

As dawn breaks, astronauts aboard the space shuttle **Atlantis** *view the earth's atmosphere, a gaseous shell extending from the surface to space. Towering clouds of thunderstorms stretch above the horizon, their flat tops, or "anvils," designating the upper limit of the troposphere—the lowermost part of the atmosphere, where weather primarily occurs.*

the tropics to nearly 6 miles in the Arctic, is the region of the atmosphere where weather occurs. It dissolves progressively into the less-dense and higher stratosphere, mesosphere, thermosphere, and exosphere, which merges into space.

As for solid earth, the planet's interior has an inner core of solid nickel and iron and an outer core of liquid nickel and iron, above which is a thick semi-molten mantle and a thin crust. That crust, made up of 16 major crustal plates, thins to only 3 miles in the deepest ocean, but it can be more than 30 miles thick under continental mountain ranges.

Constantly shifting, the earth's 16 major crustal plates have collided to build mountains or subducted to form deep-ocean trenches— from the towering 29,028-foot Mount Everest in the Himalaya, the earth's tallest peak, to the Mariana Trench, its deepest point, 35,810 feet below the surface of the Pacific Ocean.

Much of the earth remains invisible to human beings. Yet using our minds and tools, we have reached inward and outward. Only during the 20th century, and largely during the past 50 years, have humans made much progress clarifying the mystery of the earth's origin and inner structure. We have explored the earth's crust and deepest seafloor with seis-mological devices, side-scan sonar, and other imaging techniques—as well as with drills to extract rock and sediment cores. We are beginning to understand more about the unique planet that we call our home.

"Until modern times we humans shared the perspective of a tiny mayfly fluttering about a giant sequoia," geologist Robert D. Ballard once wrote. "The mayfly, with an adult life measured in hours, does not perceive the 2,000-year-old tree to be alive. After all, the insect has lived there all its life and the tree has never moved.

"We, scurrying around on an Earth that has been in existence billions of years, tend to think the same thing, that our rocky globe has no life of its own. But it has, and to understand Earth's crustal activity we must understand time— long, long expanses of time."

FAULTS AND VOLCANOES

Just past 3 a.m. on February 4, 1976, an animal din swelled mysteriously in a Guatemalan rain forest. Then, at 3:02 a.m., the ground at Gualán began to shake at the epicenter of a massive earthquake. Traveling through the valley of the Río Motagua, along a 40-million-year-old, 150-mile-long strike-slip fault, the 30-second shock was felt from Costa Rica to Mexico. Running east to west, the fault tore open a rift, throwing its north side westward some 5 feet.

Whole mountain slopes crumbled into landslides, and thriving villages turned into rubble tombs when adobe buildings broke apart, dropping heavy beams and raining debris on residents. According to estimates, more than 23,000 persons died, 77,000 victims suffered injuries, and at least a million people lost their homes.

How random nature often seems when it releases mammoth amounts of energy and shatters or crushes whatever lies in its path. The planet's surface usually seems so stable that a mountain hiker can hardly believe many of its towering peaks result from slow, ongoing collisions of continental and oceanic plates. Through earthquakes and volcanic eruptions, the earth reminds us that it can disrupt human lives and destroy structures.

What sense can anyone make of the deadly earthquake that struck Guatemala in 1976 or the magnitude 8.1 earthquake that shook Mexico in 1985? Some 200 miles west of Mexico City, a slab of ocean-floor crust known as the Cocos plate had suddenly plunged under the North American plate, sending seismic waves 1,000 times more powerful than the Hiroshima bomb racing at 25,000 miles an hour toward the city of 18 million people. So potent was the energy released by the quake that nearly 7,000 structures were destroyed or damaged, and an estimated 9,500 people were killed. Even in Texas, skyscrapers rocked.

Many scientists hope that someday they can reliably predict earthquakes and eruptions. They doubt, however, that these events can ever be controlled, a conclusion that can only exacerbate the anxieties of Japanese residents or people living along the West Coast of the United States, where recent quakes have claimed many lives. Or it may worry villagers in the shadow of volcanoes such as Colombia's Nevado del Ruiz, which erupted through an ice cap in 1985 and buried 23,000 people in rivers of hot mud.

Many of these tragedies share a common geological origin: They occurred in the so-called Ring of Fire, a 29,825-mile zone rimming the Pacific Basin and tracing the edges of enormous crustal plates. Where plates carrying seafloor plunge under plates carrying continents, friction and resulting heat melt rocks along their margins. The melted rock, or magma, then rises to the surface of the earth and spawns volcanoes. Constantly moving, the plates also generate large-magnitude earthquakes.

Such forces fueled the 1991 explosive eruption of Mount Pinatubo in the Philippines. After 600 years of dormancy, the volcano hurled 2 cubic miles of tephra into the atmosphere, sending ash, gases, and aerosols into the stratosphere. Nearly 1,000 people died from the eruption and the subsequent mudflows and disease. Had thousands of people not been evacuated, the casualties could have been as catastrophic as those in A.D 79, when Italy's Mount Vesuvius

Buried in mud and rubble, a survivor struggles to free himself in the wake of a mudflow that engulfed Armero, Colombia, on November 13, 1985, after Nevado del Ruiz erupted and melted snow and ice on its summit. So devastating was the lahar, or river of mud, which flowed down the volcano's slopes and into the Lagunilla River Valley, that some 20,000 residents perished—entombed in hot mud as they slept.

JAY DICKMAN

buried the Roman cities of Pompeii and Herculaneum. Other deadly eruptions include that of Indonesia's Tambora in 1815, which took 92,000 lives; 12,000 perished immediately and 80,000 starved when their crops and livestock were destroyed. In 1883, Krakatau, in the Sunda Strait, drowned 36,000 people in several tsunamis. Martinique's Mount Pelée, in 1902, claimed 30,000 lives when it sent a *nuée ardente*, or pyroclastic flow, into the town of St. Pierre.

Today, 500 million people live within striking distance of 550 or so active volcanoes. Presumably they have come to grips with, or at least have made peace with, the risks they face.

HIGH WATER

Like toys afloat on a vast lake, many cars, trucks, and houses in the northwestern United States drifted to ruin in January 1997. Residents were stunned when rivers of mud and other debris poured down slopes to engulf homes, and floodwaters washed away highways and railroad tracks.

Rain began the day after Christmas 1996, and by mid-January as much as 40 inches of it had drenched parts of California, Nevada, Idaho, Oregon, and Washington. Floods claimed three dozen lives and caused 1.5 billion

Floodwaters flow over a roadway at Lake Don Pedro, California, on January 3, 1997, when a dam spillway yielded to five days of torrential rain. From the day after Christmas 1996 to mid-January 1997, some 40 inches of rainfall flooded states in the Pacific Northwest, forcing more than 125,000 California residents to evacuate their homes.

dollars in property damage. In California, where the weather destroyed more than 2,000 homes and forced 125,000 people to evacuate, Governor Pete Wilson deemed the storm damage "the worst ever to hit the state."

High water from floods repeatedly proves the most lethal of weather's manifestations. In the United States, floods claim more lives annually—about 140—than lightning, which on average kills 93, or tornadoes and hurricanes, which take about 80 lives each. Flash floods borne of thunderstorms can come and go in minutes, killing without warning. Some waters rise slowly, following the seasonal cycles of monsoons or other prolonged periods of rain. Eventually, rain-swollen rivers and tributaries overflow their banks and inundate adjacent fields.

In the Great Flood of 1993, months of rain saturated a large part of the midwestern U.S.; the runoff filled the upper Mississippi River basin with floodwaters. From Wisconsin and the Dakotas to Kentucky and Indiana, the rains swelled rivers and soaked the land. On June 20, the first levee failed. In Minnesota, Kansas, Nebraska, Iowa, Missouri, and Illinois, muddy water gushed over riverbanks and poured into fields, streets, highways, and homes, coating everything with stinking mud and ooze. By late July, more than 23 million acres in 9 states north of Cairo, Illinois, were underwater. Some 50 people died, and damage estimates ranged from 15 to 21 billion dollars.

In some regions of the world, cyclical flooding is a fact of life. Consider Bangladesh, where snowmelt from the Himalaya and excessive rainfall regularly fill rivers to overflowing. Floods annually inundate a fifth of the country.

Monsoons often take heavy tolls of human life. During 1995, for example, these seasonal torrents claimed 585 lives in Bangladesh, 451 in Pakistan, and 183 in Nepal. For people living near rivers such as the Nile, Huang, Yangtze, and Brahmaputra, regular flooding is a mixed blessing. Although annual flooding covers farmlands with nutrient-rich waters, those waters can swell beyond control, sweeping away plants, animals, and people. Even relatively benign waterways have periodic problems. River ice often jams the Danube River during spring thaws, causing devastating floods like the ones in 1342, 1402, 1501, and 1830.

Europe's great cities sometimes find themselves deluged with heavy rainfall, snowmelt, or mountain runoff. Paris, for example, suffered terrible flooding during 1658 and 1910. In 1099 and 1953 storms besieged England, Belgium, and the Netherlands, inundating their coasts.

Rainy spells again swamped much of Europe in the 1990s. In southern France's Ouvèze River Valley, a 49-foot-high wall of water swept through ancient Vaison-la-Romaine in a flash flood that killed 38 people in 1992. The next year, the Meuse River overflowed, causing residents of Charleville-Mézières to swim from their homes during their "flood of the century." In 1995, the Meuse topped its previous watermark and flooded residents again. That same year powerful storms also brought floods to the United Kingdom, Switzerland, Germany, Belgium, the Netherlands, and Norway.

The banks of rivers and shores of oceans are choice spots, rich in natural resources and beauty, but those desirable locations can exact a high price in property damage and loss of life.

POWERFUL WINDS

Storm watchers knew that May 18, 1995, could be an interesting day. Warm, moist, unstable air was moving northward across the midwestern United States. Winds at low levels of the atmosphere were moving rapidly northward from the Gulf of Mexico toward a strong cold front, which extended from Michigan through Illinois and Missouri to a strengthening low-pressure system in southwestern Kansas. The cold front collided with the energy-rich air from the Gulf, causing clouds to boil high into the atmosphere.

That day, 86 tornadoes in 4 states killed 4 people and injured 161. The twister spree started in Kentucky when 3 tornadoes blasted apart homes, flipped cars, and flattened trees before 7:45 a.m. Even more tornadoes would savage the state. At lunchtime in Illinois, 13 twisters tore up homes and businesses. In Alabama, more than 90 homes, trailers, and businesses suffered extensive tornado damage. Twenty-five twisters destroyed homes in Tennessee.

No May on record has ever seen so many tornadoes. The month spawned 391—more than 12 per day—according to meteorologist Daniel W. McCarthy of the Storm Prediction Center in Norman, Oklahoma, and Frederick P. Ostby, former director of the National Severe Storms Forecast Center in Kansas City, Missouri (renamed the Storm Prediction Center and relocated to Norman). Twisters struck 36 states, even Oregon, Idaho, California, and the District of Columbia, which hadn't spotted one since 1943. But this number also speaks to progress in warning and safety. Although 16 people died in May 1995, the fatalities fell far short of the record 163 who perished in May 1952.

Throughout North America and out in the Atlantic Ocean, 1995 proved to be a fertile year for storms. More giant storms—19 tropical, 11 of which became hurricanes—brewed up over the Atlantic than in any year since 1933.

In October, flooding caused by Hurricane Opal drowned 31 people in Guatemala and 19 in Mexico. Then the storm took 8 lives in Alabama, Georgia, and North Carolina, where its 110-mile-an-hour winds downed trees; a tornado killed one person in Florida. A month earlier, Hurricane Luis ravaged the Leeward Islands of Guadeloupe and Dominica with 140-mile-an-hour winds, then swamped parts of Puerto Rico and Newfoundland, where the ocean liner *Queen Elizabeth 2* reported 98-foot waves. In August, Hurricane Erin's 100-mile-an-hour gusts sank a gambling ship cruising 90 miles east of Florida, killing 3 crew members.

High winds have caused some of nature's most destructive rampages. In tornadoes, hurricanes, thunderstorms, or downbursts, lives have ended quickly and randomly, sometimes as a direct result of furious winds, but more often because of flying debris or storm-spawned floods. While 1995 stands out as a storm year, it remains only one of many years in which hurricanes ravaged the United States, particularly along its southeastern coast.

On August 24, 1992, Hurricane Andrew flattened homes along a 25-mile-wide path in South Florida. In its wake, the hurricane left 43 persons dead, 80,000 homes ruined, 55,000 houses damaged, and 160,000 residents homeless. Most people can only imagine the horror of Hurricane Camille in 1969 and the Florida Keys storm of 1935, which killed 256 and 408

Men watch from shore while wind, rain, and waves batter a sailboat in Corpus Christi, Texas, during Hurricane Allen's arrival in August 1980. After sweeping through the Caribbean and killing 250 people there, the monster storm, whose winds peaked at 190 miles an hour, then crossed the Gulf of Mexico and dumped 16 inches of rain on the Texas coast.

people respectively, or the hurricane that struck the Galveston area in 1900, claiming 8,000 lives.

In July 1994, when tropical depression Alberto hovered over Georgia, Alabama, and Florida, few forecasters expected that 30 people would die, mostly in flash floods. In September 1989, Hurricane Hugo pummeled the southeastern U.S. seaboard and left a bill of 4.2 billion dollars, the first storm in which property damage exceeded a billion dollars. In 1992, Hurricane Iniki, the worst storm in a century to hit Hawaii, scored a direct hit on Kauai with winds gusting to 180 miles an hour. But in 1994, Hurricane Emilia, with 160-mile-an-hour winds, lived and died harmlessly south of the Hawaiian Islands.

During this century, 31 tropical storms have each killed 25 or more U.S. residents. Since 1972, however, when Hurricane Agnes sent 122 people to their deaths, no storm has taken more than 95 lives. Careful monitoring by the National Oceanic and Atmospheric Administration and the National Weather Service has helped keep coastal residents from being caught off guard.

More than anything else, modern forecasters have eliminated the deadliest of all storm elements: the element of surprise.

23

SOLID EARTH

The earth's land surface lies deceptively still. While mountains and plains may seem to repose, their state of calm presents a mere illusion—an artifact of time that is perceived on a human scale. That apparent stability belies geologic violence on a vast scale, set in motion by deep forces that occasionally shake the ground or cause volcanoes to erupt, instantly destroying cities and lives. Over millions of years, those same forces slowly reshape the surface of the earth, widening valleys, building volcanoes, and enlarging continents inch by inch. If humans could view the earth's long history from the perspective of geologic time, they would see a dynamic planet manifesting incredible powers of creation and destruction. And if they could see inside the earth they would perhaps understand why its surface is not so still after all.

At the center of the earth is a dense core measuring some 4,350 miles across—more than half the planet's diameter. Scientists believe that the core's innermost region, called the inner core, has a temperature of about 10,000°F, which is as hot as the surface of the sun. Despite the intense heat, extraordinary pressures estimated at 3.6 million atmospheres solidify the 1,500-mile-wide inner core. Researchers think that the core is mostly made of hexagonal iron crystals aligned with the rotational axis of the earth. They also think its diameter is increasing because the lower region of the outer core, which is liquid, is slowly hardening around it.

The outer core has a temperature of about 8,500°F and consists mostly of liquid iron and nickel that flow like water in turbulent currents.

In this region, the earth's rotation and the process of convection spin out spiraling liquid pillars that are parallel to the planet's rotational axis. Interestingly, the spiraling columns—like spinning electromagnets—generate electricity, creating the earth's magnetic field. Scientists speculate that the solid inner core may anchor the magnetic field.

Sandwiched between the thick core and the planet's thin, outermost layer is the superhot mantle. Like the core, it also contains two regions: the lower and upper mantle. At the boundary between the outer core and the mantle, liquid iron and nickel meet molten rock. This region of transition is thought by many scientists to range in thickness from almost nothing to hundreds of miles. Above the boundary region lies the lower mantle, a layer of dense hot rock 1,400 miles thick.

The next layer is the upper mantle, which is some 400 miles thick. Closer to the earth's surface, this region is relatively cooler and firmer than the lower mantle. The uppermost 40 or so miles of the upper mantle, together with the planet's outermost layer, are made up of rigid shifting slabs comprising the 16 major and several minor tectonic plates. The planet's outermost layer, called the crust, may measure only about 5 miles thick in ocean-floor areas and as much as 30 miles thick in regions of high continental mountains.

Within the mantle, great volumes of superhot magma slowly rise and displace masses of colder rock, which sink, warm, melt, and eventually rise again in a renewal process that releases an estimated 80 percent of the internal heat radiated by the earth. This continual cycle

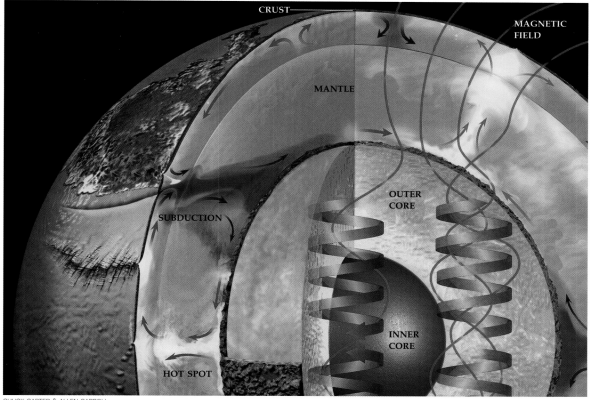

CRUST

MAGNETIC FIELD

MANTLE

OUTER CORE

SUBDUCTION

INNER CORE

HOT SPOT

Earth consists of concentric layers. A liquid iron-and-nickel outer core, about 1,400 miles thick, encases a solid inner core made mostly of iron and measuring some 1,500 miles in diameter. At the boundary between the outer core and the mantle, molten rock meets core metal, giving rise to a dense 1,400-mile-thick lower mantle. A 400-mile-thick upper mantle lies beneath the earth's crust.

EARTH'S SPINNING CORE:

■ Earth's innermost core may spin faster than the rest of the planet, gaining as much as one extra rotation every 400 years, according to recent research by geoscientists at Columbia University's Lamont-Doherty Earth Observatory. The researchers made this intriguing discovery while measuring the speed of seismic waves traveling through the earth's core.

■ Scientists know that when seismic waves are transmitted along the axis of preferred orientation of the core's iron crystals, they travel faster than when they travel at angles to the axes of iron crystals. Tracking the travel time of seismic waves from tremors in the South Sandwich Islands, near South America, to a measuring station in College, Alaska, the researchers noticed a discrepancy. Waves bypassing the inner core traveled just as fast in 1967 as they did in 1995, whereas waves moving through the core traveled 0.3 seconds faster in 1995 than they did in 1967.

■ That subtle time difference prompted the scientists to conclude that the inner core had swung toward alignment with the South Sandwich Islands and Alaska at a rate of roughly 1.1 degrees per year. Based on that observation, plus others, the team calculated the rate of rotation of the earth's inner core.

■ Geophysicists hope they can better "see" what is happening magnetically inside the planet's core, thus improving their understanding of the mysterious, perplexing magnetic field.

generates convection cells. But their pattern of movement in the mantle remains unclear. Some scientists think that the convection process within the upper and lower mantle regions is separate. If that is the case, then little or no material is exchanged. Other researchers believe that currents move through both parts of the mantle, looping from one region to the next. New studies appear to show evidence for both of these patterns.

Rising from within the mantle, convection cells provide a continuous source of molten rock, which moves up into the overlying crust and fills the magma chambers beneath volcanoes. In places such as the Ring of Fire, where volcanic islands and continental mountain chains trace subduction zones, pressure builds within the chambers and causes some volcanoes to erupt explosively, sending enormous clouds of ash and gases into the sky.

In places where tectonic plates are moving over focused hot spots of rising magma, long chains of volcanoes can form, erupt lava onto the seafloor, and produce new volcanic islands one right after the other. In time, the older volcanoes in the chain move away from the hot spot and gradually become extinct. Then the forces of weathering and erosion slowly wear them away, and they eventually sink below the surface of the ocean.

When a tectonic plate carrying relatively light continental crust—such as the North American plate—collides with a plate of denser, heavier oceanic crust—such as the Pacific plate—the edge of the oceanic slab subducts and plunges below the continental plate, triggering earthquakes as well as producing volcanic activity.

After the formation of the planet nearly 4.7 billion years ago, gravitational force, tons of impacting space debris, and internal radioactivity heated the rocky mass circling the sun. Temperatures rose during the next billion years, and iron and nickel contained in the mass began to liquefy, thus triggering the separation of the core and mantle. As the iron and nickel melted, they slowly sank to the earth's center and were compressed by gravity, raising the internal temperature about 3600°F. Ultimately, the interior of the earth separated into the three major layers: core, mantle, and crust.

Granites and basalts mostly make up the earth's rocky crust, while peridotite constitutes much of the molten mantle. Granites predominate on the continents. Basalts make up much of the seafloor and mid-ocean ridges; they also cover large areas on the continents, such as the Deccan Plateau in India. Concentrated primarily in the crust today, radioactive elements—including uranium, thorium, and potassium—

PAUL J. TACKLEY, UCLA

were found throughout the earth during its early history. These elements produced a steadily accumulating heat that helped raise internal temperatures.

Except for the sun, the planet's greatest source of heat is its interior, which conducts a hundred billion billion calories of energy to the

26

Supercomputers show where molten rock (red) rises and cold rock (blue) sinks in the mantle. In the blue three-dimensional image (opposite), oceanic plates plunge into the mantle, where they are absorbed. In the red image, superhot rock nears the surface in mushroom-shaped plumes. Seismic tomography of the Pacific region (below) locates cold slabs that descended long ago, as well as hot rock rising from the lower mantle.

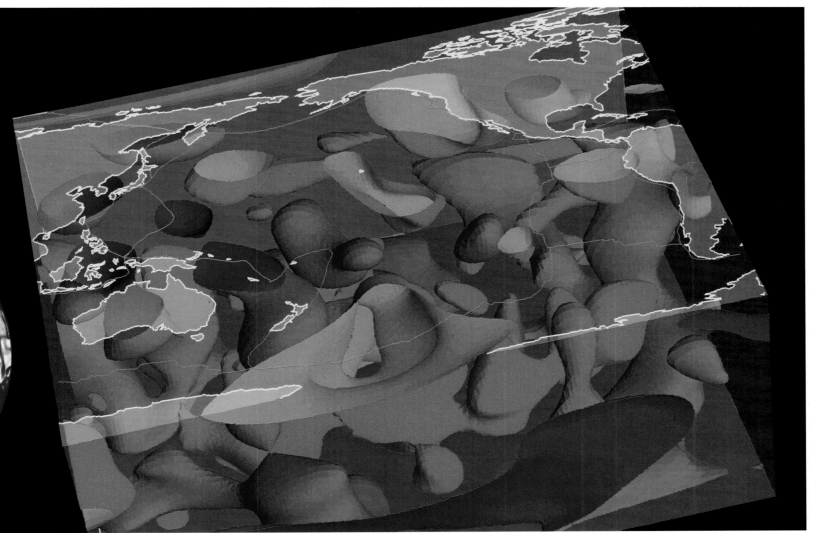

surface each year—or 1.5 microcalories per .155 square inch of the earth's surface each second. In fact, so much energy radiates from beneath the planet's surface that it could satisfy all human energy needs three times over. Much of the energy that reaches the land surface, however, radiates into space.

Research scientists drilling holes as much as 5 miles into the crust find that rock temperature increases roughly 37°F per 320 feet. Even on the deepest seafloor, rock cooled by bottom water remains slightly above freezing.

Earth's inner heat ensures that its surface will remain active—and habitable.

SHIFTING PLATES

The earth's crust is composed of a series of moving plates that converge, diverge, or slide past one another, occasionally releasing energy through faulting, which generates seismic waves. The plates move just a few inches a year—about as fast as a person's fingernails grow—and this movement is called plate tectonics. The word "tectonics" is derived from the Greek word for builder. Roughly 16 major and several minor plates make up the planet's rocky outer layer, or lithosphere, and they range in size from a few hundred miles to several thousand miles across. Among the largest ones are the Pacific and Antarctic plates.

Thinner under ocean basins and thicker under continents, the shifting plates that make up the rigid lithosphere rest atop a layer of semi-molten rock called the asthenosphere. Below the asthenosphere lies the mesosphere, a thick layer of semi-solid rock. Scientists have yet to concur on exactly what causes plate movement, but they agree that heat within the earth's mantle and outer core supplies the energy for it.

A quick glance at the earth's surface reveals two obvious features—land and water. This division of landmasses and oceans owes its existence to differences in the thickness and composition of the planet's two types of crust. Continental crust, which is mostly composed of relatively lightweight granite, is some 18 to 30 miles thick. The much denser basalt of oceanic crust measures only 5 or 6 miles in thickness. Because of their greater buoyancy, the continents "float" higher on the mantle than the ocean floors do, and this fact accounts for the differences between the two levels of the earth's surface.

As the plates travel, they interact along their edges. And after having studied how the plates move in relation to one another, scientists have recognized and defined three types of boundaries: convergent, divergent, and transform-fault zones. All told, scientists estimate that as much as 80 percent of all earthquake and volcanic activity happens on or near these plate boundaries. They also say that about 80 percent of all volcanism occurs in the oceans.

Where plates diverge, or move apart, rifts provide openings that let molten rock flow out of the earth's upper mantle and crust. This "spreading" process, usually occurring along the Mid-Ocean Ridge, creates new crust as part of an ongoing process of ocean-floor renewal. Along these boundaries, the outpouring of lava tends to create volcanoes. In the Afar region of East Africa, for example, volcanoes rise along a rift, or graben, made by diverging plates. From Uganda to Malawi, a group of lakes fills many of the grabens formed by tectonic spreading. In the North Atlantic Ocean, faults and grabens reveal a major spreading zone across the volcanic island of Iceland, which straddles a divergent boundary.

Convergent boundaries, in contrast, form where two plates slowly collide. Typically, one plate rides up onto the leading edge of the other one; the heavier oceanic plate subducts into the mantle, where it becomes molten and is reabsorbed. Such boundaries are usually marked by deep submarine trenches and lines of volcanoes that are generated as a result of subduction. The Aleutian Islands, for example, formed because

Deformed layers of an exposed southern section of the San Andreas Fault show huge folds in solid rock. Revealed during the construction of State Highway 14, the rocks in this road cut near Palmdale, California, lie only 200 yards from the active fault and a mere 45 miles north of Los Angeles. Shifting tectonic plates build up temperatures so high and pressures so intense that they literally crumple rock layers—tilting, folding, and faulting them over millions of years.

of volcanic activity along a subduction zone on the northern edge of the Aleutian Trench. Farther south, the Juan de Fuca plate, a remnant of an earlier and larger Farallon plate, is plunging under the coast of Oregon and Washington on the North American plate. Someday it will disappear completely, but for now it generates volcanic activity in the Cascade Range.

Where two plates slide past each other, a transform fault forms. In California, the San Andreas Fault system traces a long stretch of the boundary between the North American and Pacific plates. For ten million years, the Pacific plate has been sliding past the continent at an average rate of two inches a year. Because the plates move laterally along this kind of boundary, no new crust is created. Large earthquakes, however, are frequently produced.

Tectonic activity also may occur when a plate travels across a stationary hot spot in the earth's mantle. In such cases magma rises into the crust, spews out as lava, and builds giant shield volcanoes like the ones in Hawaii. Sometimes the magma warms groundwater, producing geysers and hot springs like the

ones in Yellowstone National Park and Iceland.

Using bathymetric surveys, Fathometers, side-scan sonar, and core samples taken from the seafloor, scientists have investigated parts of the Mid-Ocean Ridge, a 46,600-mile-long undersea mountain range that extends around the globe. Along the ridge's crest runs a deep central rift where oceanic plates are separating. Magma moves from the upper mantle into the overlying crust, causing lava to flow into the cold ocean waters here and form new crust. Scientists estimate that for more than 150 million years seafloor spreading along the ridge has been forming the floor of the Atlantic Ocean. As new crust is made, old crust descends into the mantle along subduction zones and is then melted and recycled. Although new crust is constantly being created, the earth stays the same size because older crustal material descends into the mantle at the same rate.

In some places, oceanic plates converge with continental plates. Along the western coasts of North and South America, the lighter plates supporting the American continents are riding up on several plates that form the floor of the eastern Pacific Ocean, causing the Pacific plates to plunge beneath the continents and forming submarine trenches on the western margins of Central and South America. The plunging plates generate heat that gives rise to chains of volcanoes. In the case of the Nazca plate plunging beneath the South American plate, magma reaching the surface has built Chile's spectacular Andean peaks.

When two continental plates collide, a vast mountain range may result. The Himalaya began to rise several million years ago when the northward-migrating Indian plate, carrying continental crust, collided with and slipped under the Eurasian plate's continental crust. Today the snowcapped peaks of India and Pakistan's Karakoram Range bear testament to the power of continental collisions.

At times, material on colliding plates may enlarge a continental landmass. This may happen when a plunging plate carrying a seamount or other landmass reaches a trench; there the chunks of crust fuse with the overlying plate. Through this process, called accretion, pieces of continental crust become packed together like puzzle pieces, spawning a variety of accretionary landscapes commonly seen, for instance, along the western coast of North America, particularly in Alaska and California.

Many geologists did not believe that the continents were in motion until the late 1960s. The theory, however, dates back to the early 20th century. In 1912, the German meteorologist Alfred Wegener put forth a theory of continental drift. He postulated that a single great landmass, called Pangaea, dominated the earth during the Upper Paleozoic and Lower Mesozoic eras, between 300 and 200 million years ago. This supercontinent, he proposed, broke apart and its pieces migrated to assume the shapes and positions of today's continents.

Wegener based his theory on the fit of continental coastlines such as western Africa and eastern South America, on correlation of fossils and past climates in both regions, and on new information learned from mapping similar rocks and minerals in mountain ranges on both continents. While many skeptics dismissed Wegener's ideas, a few found them intriguing enough to continue the debate and study the possibility that continents actually do move. Supporting evidence slowly accumulated. Not only do the opposing Atlantic shorelines match, but the tectonics and structure of mountain ranges in eastern North America and northern Europe also match well. Glacial deposits were found in equatorial regions, while deposits of coal—the remnants of abundant vegetation and a warmer climate—were found in Antarctica.

Some of the earliest persuasive evidence came from fossils. *Lystrosaurus,* an ancient reptile whose fossils were first found in India and South Africa, also was discovered in Antarctica. Scientists concluded that the remains of a tropical reptile could have found their way to Antarctica only if the now frozen continent had once stood in a warm climate; similar rock formations also confirmed that it had once been attached to India and Africa. In the Sahara, fossil shells of now extinct marine creatures—ammonites—were discovered in the Middle Devonian to Upper Permian rock formations.

Researchers also found that rock formations

Reaching from Colorado to Texas, the ever widening Rio Grande Rift and its namesake river zigzag across northern New Mexico. Plate movement and volcanic activity continue to enlarge the down-faulted rift, or graben, which already cuts into the earth six times as deeply as the Grand Canyon. Sediments and basalt fill in and conceal much of the crustal split from residents living nearby.

on South America's eastern coast matched rocks on Africa's western coast, both dating back some 550 million years to the Cambrian period. And geologists discovered that volcanic rocks, which can be used to document changes in the earth's magnetic field, revealed that the north magnetic pole has migrated since Precambrian time.

In 1987, scientists used very long baseline interferometry to prove Hawaii's movement toward Japan at some three inches per year, as well as the slow separation of North America and Europe at about one inch per year. While the Atlantic Ocean slowly widens, researchers are watching the Pacific Ocean gradually shrink. Perhaps western California will eventually separate from the rest of North America and become an island.

UNDERWATER RIFTS

Thousands of feet below the ocean's surface, cloaked in inky blackness and subjected to extreme pressures, life manages to flourish. Somehow, organisms living next to undersea vents thrive in a world human beings can only fleetingly visit and barely understand.

In 1977, at a site 235 miles northeast of the Galápagos Islands, scientists found proof that life can exist around hydrothermal vents. In the headlights of their submersible *Alvin*, some 8,000 feet deep, geochemists watched two-foot-long tube worms clustered near lava vents that spewed mineral-rich, bacteria-laden jets of boiling water. Nearby, foot-long white clams and yellow-brown mussels lay in lava fissures, while blind white crabs scavenged for food.

As early as the mid-1960s, researchers predicted that red-hot lava from seafloor spreading might produce hydrothermal vents in the deep ocean. Yet, in such a cold and lightless world, no one expected to see much life. Until recently, scientists thought that the ocean's deepest spots, long viewed as largely lifeless deserts, could not support a specialized ecosystem. Now researchers realize that volcanic heat supplies energy necessary to sustain deep-ocean, self-sufficient animal communities. Unlike surface life, which derives energy from the sun, these deep-bottom dwellers draw energy from the earth's interior.

As the plates covering the earth's surface move apart, magma is intruded into the Mid-Ocean Ridge; lava flows from fissures along the ridge axis, forming new crust. Seawater seeps down through fissures in the crust, descends several miles, and under enormous heat and pressure leaches minerals from rock. The water spews from vents in scalding jets, and rapid cooling precipitates deposits of iron, copper, and zinc that form chimneys up to 160 feet tall. So much water circulates through the planet's Mid-Ocean Ridge that scientists believe each drop of seawater makes this passage once every ten million years.

Because most seafloor spreading occurs along the globe-girdling Mid-Ocean Ridge, hydrothermal vents teeming with life turn up most often here. While the floor of the Pacific Ocean reveals more active spreading, in recent years scientists also have observed nutrient-rich oases of life along segments of the Mid-Atlantic Ridge. Residents there include shrimp bearing anatomical structures that act as light-collecting receptors. This finding fascinates researchers, who had long held that the deepest abyss remained pitch black, a belief supported by many eyeless animals.

After finding shrimp that are apparently adapted to the environment, researchers discovered that a faint glow—prompted by an unidentified inner-earth process—emanates from the vents. While such radiation falls well below the human retina's detection level, it can be measured by instruments. Researchers are continuing their study to see if creatures are living off this light, or using it.

The complex chemistry of thermal vents bathes local organisms in a food concentration perhaps 500 times greater than what usually exists in the deep ocean. Hydrothermal fluids rich in hydrogen sulfide, an otherwise toxic gas, are used for food by thriving vent bacteria. These bacteria form the base of the food chain.

Some of them dwell inside clams and tube worms, aiding in digestion, or serve as food for lower organisms that then become food for higher ones.

Not all vents are created the same, scientists have learned. They bear distinctive signatures and spawn a variety of ecosystems. The low-lying thermal vents discovered along the Galápagos Rift, for example, barely resemble the tall chimneys or Hanging Gardens seen at 21°N, on the East Pacific Rise. Here, ominous "black smokers"—almost like industrial smokestacks—emit black, metal-rich clouds as hot as 662°F. The vents' biodiversity varies too.

At 21°N, the football-field-size sprawl of gigantic white clams, known as Clam Acres, contrasts notably with the snail-coated chimneys of Lau Basin, near Fiji.

Because of shifting plates, flowing lava, and constant changes on the seafloor, thermal vents are thought to exist only for a limited period at the same location. When a vent ceases to function, the animals that depend on it either die or relocate. In 1979, for example, biologists found species at 21°N that had been seen on Galápagos vents 2,100 miles away, suggesting that vent animals disperse larvae great distances to ensure survival near new vents. In contrast, when researchers spotted new species of inch-long clams and smaller tube worms at the Juan de Fuca Ridge off the coast of Washington and Oregon, they realized that some communities may evolve independently.

Giant tube worms—some 10 to 12 feet tall—thrive beside an undersea "chimney" at a vent in the Galápagos Rift. Hot, mineral-rich geysers feed colonies of these organisms, whose bodies contain bacteria that metabolize nutrients from hydrogen sulfide in the seawater and make food for their hosts.

Life's ability to thrive in such seemingly inhospitable environments has caused many biologists to reevaluate the idea that life had to evolve in a solar-dependent environment. Some biochemists now argue that life may have originated in bacteria-colonizing thermal vents, an unusual but plausible alternative to prevalent theories on the origin of life. While researchers have explored less than one percent of the earth's spreading seafloor, they believe that life-forms probably flourish wherever volcanic activity fuels hydrothermal vents.

PLANETARY CRUST

Constantly, steadily, and sometimes dramatically, the sculpturing forces of nature reshape the surface of the planet, creating new landforms and modifying the older ones.

Wind, water, and glacier ice are nature's favorite sculpturing tools, and they are used every day to carry out the ongoing processes of weathering and erosion. Weathering is the slow physical and chemical disintegration of solid rock by water, ice, acids, plants, and fluctuations of temperature. Erosion describes the group of processes that mechanically transport soil, rock, and other geologic materials by such means as running water, wind, glacier ice, waves, and currents. Water, even when gentle, can eventually wear down any exposed rock and then carry away its fragments.

Ice, in the form of great glaciers, has periodically advanced and retreated over the North American and Eurasian continents, altering old landscapes and creating new ones. Just in the past one to two million years, the earth has experienced several ice ages.

Some 18,000 years ago, during the height of the most recent ice age, vast sheets of ice spread over northern Europe and much of Canada and the northern United States, covering perhaps one-third of the planet's land surface. From Canada the Laurentide ice sheet flowed southward, eroding the landscape and leaving huge deposits of till as ground and terminal moraines. In time, glacial meltwater filled depressions left by the ice, creating numerous lakes across the northern U.S. from the Rocky Mountains to New England. In western New York State, streams within the Laurentide ice

PAUL CHESLEY

34

At 15,771 feet, Mont Blanc towers over the village of Le Mont in France's Chamonix Valley. The mountain range's craggy facades testify to its young age and the power of glaciers that sharpened peaks and deepened valleys over thousands of years in the Alps, Andes, and Himalaya. So magnificent are Mont Blanc's vistas that French novelist Victor Hugo called the peak the "white shepherd of stormy mountains."

sheet eroded U-shaped basins and troughs that, when filled by water, became the well-known Finger Lakes. Paths followed by the glaciers have been preserved in rock: Deep scratches in bedrock—or striations—show the direction of the ice flow. Some of the earth's most spectacular landscapes owe their existence to the erosional work of ice. Among them are Wyoming's Teton Range and California's Yosemite Valley, the Alps' Matterhorn, which graces the border between Switzerland and Italy, and Patagonia in South America.

As the earth's climate ameliorated, most of the great ice sheets melted, leaving behind enormous deposits of sand, clay, gravel, and boulders. Sometimes clogging valleys or spreading into long ridges of land, these mounds of debris, or moraines, dammed rivers to spawn lakes. They became peninsulas such as Cape Cod, Massachusetts, or formed islands like Long Island, New York. The ice sheets transported boulders as large as houses, and when the ice retreated, the boulders were dropped and left behind. Because they are different in composition from the surface upon which they rest, these boulders are called erratics.

The Columbia Glacier, a valley glacier in Canada's Rocky Mountains, reveals ongoing glacial landscaping. Flowing downhill from bowl-shaped mountain hollows known as cirques, valley or alpine glaciers slowly erode V-shaped valleys into U-shaped ones. When several mountain glaciers descend, they can sculpture sharp divides or jagged, knife-edged ridges called arêtes between the U-shaped valleys. Ice streams flowing in an ice sheet can erode steep-sided valleys beneath a glacier;

these depressions become fjords when the ice sheet melts and the sea invades them. Today, glaciers still cover about one-tenth of the earth's crust. Among them are the thick ice sheets in Antarctica and Greenland, which remain from the most recent ice age, and the ice that covers polar islands and caps high mountains, even at the Equator.

Nature's hand is not always heavy, however. Its touch may be lighter—in the form of wind and water—but it is no less effective in shaping the land. No rock can long withstand the daily effects of weathering. Over eons, rainfall and the unending cycles of freezing and thawing have eroded the once towering peaks of North America's Appalachian Mountains, carved the Grand Canyon's awesome troughs, and fashioned rocky spires in Utah's Bryce Canyon. Exposed surfaces wear away, sharp edges soften, and rocks crumble, revealing, for example, the magnificent sandstone cliffs of Canyonlands National Park in Utah or the buttes and mesas of Monument Valley, in southern Utah and northern Arizona.

Water seeps between mineral grains and fills rocky crevices, which grow deeper and wider when the water freezes, expands, and acts like a wedge to enlarge them. Even roots of a tree growing in a crevice can split the rock. Acidic solutions from rainfall and biological decay eat away at minerals in solid rock. Dissolution by water of limestone formations creates great networks of underground caverns, such as the vast labyrinth of Mammoth Cave, in Kentucky, or New Mexico's Carlsbad Caverns. In Arizona, flowing water has eroded spectacularly shaped canyons.

Turret Arch, one of 300 such formations in Utah's Arches National Park, results from the differential weathering of sandstone. Water trickling beneath a layer of softer stone can dissolve it away, slowly creating a cave-like opening. Over time, the opening grows deeper and wider. Erosion by water and wind rounds its sharp edges into an arch shape.

The combined effects of weathering and erosion would eventually level all landforms, thus leaving a featureless plain, if the earth's crustal plates did not move. Even at the planet's surface, weathering and erosion leave their mark in new soil created by burrowing animals or decaying carcasses, spreading tree roots, dissolving bedrock, and seeping natural acids made by atmospheric processes. Rocks containing iron, for example, may simply rust and crumble through natural oxidative processes.

When rock yields to weathering, its fragments may move. That movement, or erosion, is helped by wind, water, and ice, which transport sediments—ranging from boulders and pebbles to sand, silt, and clay—from their place of origin to their new place of deposition. Water plays a critical role in erosion and movement of sediment. Muddy waters, carrying suspended soil and dissolved minerals, signal ongoing erosive processes. Flash floods caused by sudden, heavy rainfall can scour land surfaces, transporting large amounts of silt, clay, and other sediments from some areas and eventually depositing the suspended particles in other areas. Utah's San Juan River normally carries so much mineral sediment that its waters display a salmon tint.

Water flowing down mountainsides can carry gravel, sand, and clay and move them to a new resting place in a valley or plain below. Such deposits of alluvium tend to spread out into fan-shaped beds, or alluvial fans. Rock formations are not homogeneous. Water can produce differential erosion of weak and hard rock, creating openings in rock that are slowly enlarged as an arch, for example. Some 300 natural arches grace Utah's Arches National Park.

Where sediment-rich streams flow across the land, they can establish stream networks that will swell into river channels in valleys. Water that merges into a river eventually finds its way to a larger body of water. At that juncture—the river's mouth—large deposits of rock, sand, and clay will create a delta, a landform that has proved over the centuries to supply especially fertile soil for agriculture. Indeed, when rivers overflow their banks, they can cover their floodplains with sediments and thus replenish farmlands. Inhabitants along the banks of the Nile River in Egypt, for example, long used the natural flood cycles to their farming advantage.

Eventually, all river-borne sediments end up in the ocean. While a particular rock or soil particle may take thousands of years to be transported from mountains to ocean, each year the world's rivers deposit millions of tons of sediment into the sea. The waves of the oceans lap and pound at continental shores, shaping, molding, and eroding coastlines while grinding boulders and pebbles into sand. Pounding against rough-hewn seaside cliffs, the waves

may erode holes and create caves, eventually shaping them into arches that collapse to form sea stacks like the ones along the Oregon coast or Canada's Bay of Fundy. If they too crumble in time, their remnants turn to sand, which waves then transport to nearby shores.

Particles of sand transported by waves and currents along the coast are often removed from one beach and deposited on another. Along the southern coast of New York's Long Island, for example, once broad islands are becoming narrower—particularly the barrier beach of Fire Island, which has been overwashed by several hurricanes. On the islands that make up North Carolina's Outer Banks, waves erode the fragile

The Grand Canyon of the Yellowstone River displays spectacular falls. Water gushing over its Lower Falls plummets 308 feet into a great gorge. Slightly upstream, the water of the Upper Falls plunges 109 feet into misty froth. The falls show where volcanic emanations and flowing water slowly transformed the canyon's hard walls of rhyolite—a lava rock layering Yellowstone's caldera—from gray-brown hues into pinnacles streaked with red, white, and yellow. Referring to the Yellowstone River's extraordinary beauty, English writer Rudyard Kipling once remarked: "Now I know what it is to sit enthroned amid the clouds of sunset."

barrier beaches. Hurricane waves and storm surges have also assaulted this long stretch of the East Coast. In 1870, the Cape Hatteras Lighthouse stood 3,200 feet from the shoreline. Today, less than 192 feet of sand separate the lighthouse from the curling surf.

Along the outer beach of Cape Cod, Massachusetts, the coastal cliffs are being eroded at the rate of about three feet per year. The sand is being transported to the north, extending the coast at Province Lands, as well as to the south, toward Monomoy Island.

Because of the melting of glaciers and the warming of the upper oceans, sea level is rising about a tenth of an inch per year. It is estimated that 70 percent of the world's easily erodible coastlines are retreating in response to this rise in sea level.

Wind, as well as water, can transport sand, dust, soil, and volcanic ash and carry them to far-off locations. Wind builds dunes of sand and molds the material into shapes such as ridges, curves, crescents, and stars. Throughout the middle latitudes of the United States, China, Russia, and Argentina, for example, winds have built up deposits of fine silt and sand, or loess, culled from clay and silt dropped by streams from melting glaciers. Sand-laden gusts, too, can erode boulders or turn cliffs into sculptured masterpieces. In the Baja California desert, for example, boulders have been worn smooth by sand and wind.

Unlike water, which must submit to the downward pull of gravity, wind can create deposits of sand and then pile them into rolling dunes. Such dunes grace many shorelines, including the beaches along two of North America's Great Lakes. They also form vast sand seas that stretch across parts of the Middle East and North Africa. A stealthy conqueror of lands, dunes moved by winds envelop vegetation and bury the remnants of previous inhabitants. In a matter of decades, a rooftop poking through the sand may stand as the only evidence of a former desert dwelling. Inhabitants of barrier islands prize protective dunes, struggling to stem their erosion by planting beach grasses and other plants that will help stabilize and anchor the sand.

The ability of wind to pick up loose sediment, especially after a prolonged drought, can lead to blinding dust storms that loft tons of sand and dust high into the air and resettle the material thousands of miles away. With little vegetation to block it, wind can sweep through desert areas, lift sand and dust, and then carry the particles across continents. Dust from Africa crosses the Atlantic Ocean and often changes the color of Florida's sunsets.

Unfortunately, human activities can accelerate wind's erosive might. Poor farming practices and deforestation can leave topsoil exposed, only to be lofted and whisked away from where it is needed. In a severe drought in 1936, for example, much of the topsoil of the southern Great Plains of the United States blew away, leaving a ravaged land behind. Overgrazing and tree cutting has damaged grasslands in the West African Sahel, hastening expansion of desert areas. To combat such problems in China's Tengger Desert, where sand dunes are moving, researchers have assembled a gridlike network of straw fences to anchor the otherwise drifting sands.

RAPID LANDSCAPING

Sometimes nature strikes in a flash. In a matter of minutes a landscape can be transformed from a pleasant vista to an inferno or a tomb. Volcanic activity and earthquakes often trigger avalanches and mudslides, some minor, some deadly. In the worst cases, tons of rock, snow, or mud can unexpectedly flow into populated areas below.

On November 13, 1985, barely an hour before midnight, a volcano in Colombia produced a minor eruption that melted part of its ice cap. A mudflow, or lahar, roared down the slopes of Nevado del Ruiz and onto the sleeping residents of Armero, an agricultural center 30 miles to the east. Other low-lying towns also were buried. In places, victims were entombed in mud nearly 12 feet deep. About 20,000 of Armero's residents died; 1,000 people who lived in Chinchina on Nevado del Ruiz's western slope also perished.

Volcanoes erupt lava, tephra, gases, and aerosols. They can also produce lahars, or mudflows, and roiling clouds (*nuées ardentes*) of glowing ash. Flowing lava can cover the land or build distinct islands such as Hawaii or Iceland. Explosive eruptions, on the other hand, can cover entire cities with lava and ash, as did the A.D. 79 eruption of Mount Vesuvius, which destroyed Pompeii, Stabiae, and Herculaneum.

Ten percent of the world's active volcanoes are in Japan, and many residents there rely on barriers designed to protect them from the inevitable lahars. On the slopes of Hokkaido's Mount Usu, for example, check dams built across natural drainages, as well as concrete walls lining the sides of gullies, can slow and direct the mudflows.

For decades, actor Charles Laughton's former home (opposite) sat atop cliffs in Malibu, California. Then, in January 1994, an earthquake damaged it. Heavy rains followed, and a mudslide carried away part of the house. Rainfall can saturate soil and loose rocks on slopes and in road cuts, causing them to suddenly move downhill as landslides. In Oregon, a rockslide on Highway 101 wrecked a car (below).

GARY BRAASCH/TONY STONE IMAGES

Like lahars, avalanches—whether landslides of rock and soil, or snowslides of snow and ice—have crushed towns and altered landscapes. In 1964, an earthquake in Alaska opened huge fissures in the Turnagain Arm area of Anchorage; many homes slid into the bay. In the Peruvian Andes in 1970, a massive avalanche entombed the town of Yungay in rock and ice, leaving 18,000 dead.

In the western United States, about 100,000 snow avalanches tumble from mountainsides each year. Snowslides outpace skiers; a few plummet at more than 200 miles an hour. Landslides usually follow heavy rains, when water saturates soil, gravel, or clay supporting boulders and rocks. If rocks on a mountainside become unstable, gravity or their own weight can send them downhill.

Snowslides mostly follow the development of overloaded snowpacks that have built up unstable layers of ice and snow during periodic freezes and thaws. The tiny vibrations of skiers or an earthquake can set them off or trigger a chain reaction. Then layers of snow and ice start to slide, picking up speed and gathering more material until a major avalanche occurs.

Once moving, a slab of snowpack can fracture into rolling blocks that break into smaller pieces and rise into the air. A full-fledged avalanche can deliver a million tons of snow to a valley below. Snowslides are so common and potentially deadly, particularly in steep ranges such as the Alps and Rockies, that ski patrols and avalanche-control teams sometimes use explosives to set them off and thereby control their falls. In Switzerland, the residents of Andermatt have cultivated protective evergreens as a natural barrier above their village, while villagers of Wengen have erected fence-like steel barriers to stabilize the snow.

RAPID LANDSCAPING / YOSEMITE ROCKFALL

At 6:52 p.m. on July 10, 1996, as hikers enjoyed the late-day sun, California's Yosemite Valley echoed suddenly with the crashes of massive rock slabs hurtling down a hillside and descending into the valley below.

From their perch 2,600 feet above Happy Isles, an area having a nature center and a nearby campground, two pieces of a 495-foot-wide slab fell from a granite arch seated between Washburn and Glacier Points. The slabs slid 500 feet along a steeply inclined plane, launched into the air, plunged 1,700 feet to an underlying talus, and smacked the ground at well over 160 miles an hour—like a book dropped on a floor,

according to Gerald Wieczorek of the U.S. Geological Survey.

The air blast propelled by 68,000 tons of shattered rock flattened hundreds of pine and oak trees, bulldozing them "like bowling pins, scattering them in streams and along the popular Mist and John Muir trails," according to Maureen West of the *Phoenix Gazette*. Yosemite historian Jim Snyder writes that many people likened the blast-cleared area to "landscapes left in the wake of Mount St. Helens."

Tragically, blast-felled trees killed 20-year-old Emiliano Morales and injured several others, two seriously, while crushing the snack

42

In California's Yosemite Valley (opposite), a rescue worker and his dog search for victims following the July 10, 1996, collapse of an arch that killed one person. A computer model (below) depicts the rockfall, which began at a release point (red), slid down a slope (dark yellow), went airborne (blue), and smashed onto a talus slope (white). The air blast felled trees (gray) and blew up dust for miles (sand).

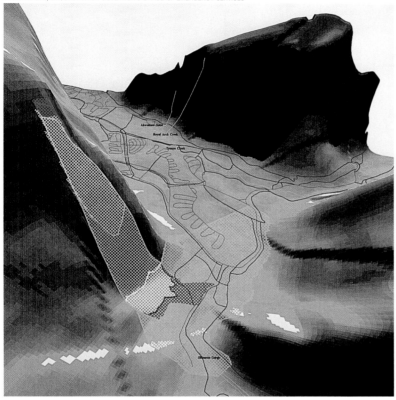

stand and damaging bridges, the nature center, trails, water mains, and signs.

The air blast lofted tons of granite dust into the air, enough to coat an estimated 50 acres, while the impact of the larger block registered 2.1 on the Richter scale. "It felt like an earthquake and sounded like cannon-fire," reported Bill Leavengood, a Florida visitor. Jogging on nearby Vernal Falls trail, Ernie Milan worried that a jet had crashed. Then came tornado-like dust clouds—"a boiling wall of gray," said camper Roger Johnson—that whooshed in to envelop nearby land and blacken the sky. Sirens blaring, emergency vehicles plowed through

inky dust that hung for hours over the rockslide. "It's like being in a dust storm and searching for a needle in a haystack," commented John Brenner of the Sacramento Fire Department.

Although rocks often fall in Yosemite, only once before did a collapse explosively blast the air this forcefully. On March 26, 1872, an earthquake toppled a slab from Yosemite's Liberty Cap, emitting a gust that reportedly knocked a nearby hotel two inches off its foundation.

The cliff that launched the 1996 rockfall stands above Yosemite's most recent level of glaciation, which occurred between 28,000 and 17,000 years ago. The rock that failed had weathered over the past 750,000 years.

More than 400 documented Yosemite "slope-movement events" show that rockslides and falls frequent the land more often than debris flows or slumps. While storms and earthquakes provide the best recognized rockslide triggers, most of Yosemite's rockfalls reveal no known trigger. Although snowmelt, freeze-thaw cycles, and human activity also destabilize slopes, quake-triggered rockslides produce the largest volume of debris. In 1980, for instance, quakes around Mammoth Lakes, California, triggered more than 5,000 rockslides and falls, turning an estimated 35 billion cubic feet of rock into deposits of boulders, gravel, other rock fragments, and dust.

Most rockfalls take place during autumn and spring. Storms occur and saturate the cliffs, snow rapidly melts, and fluctuating temperatures cause a cycle of freezing and thawing that weakens rocks. The recent Happy Isles rockfall happened in July, however, as did a slide in the same spot on August 2, 1938.

COLLIDING CONTINENTS

ROBB KENDRICK

trange as it seems today, India once was an island continent. Part of the Indo-Australian-Antarctic plate, it drifted northward at the geologically startling rate of six to eight inches a year. Then, some 50 million years ago, two continents collided: The Indian plate met the Eurasian plate.

Today's geological evidence on land only partly tells the tale. More than 180 million years ago, when the ancient supercontinent called Pangaea started to break up, the earth's continents began to make independent journeys around the globe. Some 10 to 20 million years later, a new lithospheric-plate configuration formed, its northern edge carrying the floor of the Indian Ocean, the old subcontinent, and a sediment-covered continental shelf. Moving northward, the Indian plate's leading edge shoved beneath the Eurasian plate, ramming India's continental shelf into the continental mass known now as Eurasia.

Chains of volcanoes rose above the subducting plate and lava flowed onto the land, while granite intrusions welled up below the rising mountain mass. India's continental shelf shoved under the Eurasian plate. The colliding landmasses incorporated fragments of the old seafloor that had separated them, then moved rock and sediment layers against land now called Tibet. The Himalaya began to rise. By 28 million years ago the landmass of India was part of the continent of Asia, and sediments eroded from the new mountain range were folded, forming the foothills of the Himalaya.

The massive continental collision continues to this day, with India moving 1,240 miles into Asia, compressing China and Mongolia eastward, while moving Tibet northward and raising it three miles above sea level. The Himalaya system continues to increase in elevation as the Indian plate moves at a slightly slower rate of two inches a year, producing periodic earthquakes to relieve the stress of the underthrusting.

Some 65 million years ago, when the Indian plate moved northward, a rising magma plume, or hot spot, erupted through its continental crust. Lava spewed onto plains, piling up as much as 7,800 feet—more than one-fourth the Himalaya's height—to produce the Deccan Plateau. India's passage over the hot spot produced a line of volcanic seamounts on the Indian Ocean's floor. Réunion Island, along with its active 8,629-foot-high volcano, Piton de la Fournaise, today overrides the hot spot.

Millions of years from now, India will probably slow its northward march into Eurasia and grind slowly to a halt. By that time, today's plates will have changed shape and direction, yielding to the earth's ongoing processes that reabsorb the old and forge the new. Already we can spot slow but steady changes in existing continents, and what we are seeing is providing clues about geological processes. East Africa has begun separating from the rest of the continent and may someday collide with southeastern Asia. Should those two landmasses collide, they will likely spawn new mountain ranges and volcanoes.

Continental collisions and lava eruptions still persist and will continue—so long as the earth persists. Rifts and volcanic hot spots on the planet's seafloor tell us that the earth's interior remains hot. The Hawaiian Islands and

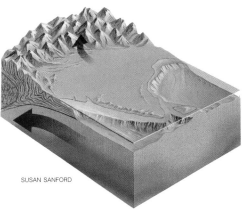

Sherpas pass ice spires (above) while trudging toward Everest, the highest peak in the world. The drawing (right) shows the northward-thrusting Indian subcontinent wedging under the Eurasian plate, triggering earthquakes while uplifting plateaus and the Himalaya.

SUSAN SANFORD

their eroded relatives, such as the Emperor Seamounts, bear testament to ongoing geological processes. After 70 million years, the northernmost volcano in the island chain stands poised to plummet into the Aleutian Trench. So the cycle continues: As plates drift, old crust subducts, melts, and becomes magma that later rises to the earth's surface and yields new volcanoes.

45

WEATHER

On a clear day from San Francisco's Golden Gate Bridge, visitors and residents alike can view the breathtaking expanse of the sapphire bay, the city's striking skyline, and the rolling hills of Marin County, California. But such vistas remain rare. More often, a dense gray fog appears and settles over the tidal waters of the bay. On some days the fog envelops the bridge so completely that the entire steel structure seems to vanish.

Dense fog engulfs San Francisco's Golden Gate Bridge. Throughout northern California's bay area, mist and fog blow in from offshore, where warm, moist, low-lying breezes cool rapidly over the Pacific Ocean's chilly currents. Such advection fogs commonly occur along the Pacific coast, drawn inland by low pressure.

While exquisitely ephemeral, such a cloud still projects a powerful illusion. Its apparent solidity, however, is actually the temporary coming together of trillions of water droplets.

Some of those droplets may have started from sea spray, while others could have once resided on a frozen peak, within a puddle on a city sidewalk, or in a child's breath. Such is the nature of water, which has the ability to evaporate, rise, and then condense into dew, frost, fog, rain, snow, or clouds. Carried by westerly winds, water vapor embarks on circuitous journeys around the globe. In some sense, weather tells the story of those movements.

The energy driving the planet's weather systems and atmospheric circulation begins in its nearest star. The sun, radiating heat and light onto the tilted, rotating earth, warms the surface of the planet unevenly. The resulting temperature differences cause tropical air masses to heat, expand, and rise while polar air masses cool, contract, and sink.

Heat from the sun daily evaporates trillions of tons of water from the earth's oceans, lakes, rivers, and land, and energy is then stored in water vapor until it is released during condensation. Warm, dry air acts like a sponge, soaking up moisture until it becomes saturated; cooling that saturated air mass—like squeezing a wet sponge—causes the water it holds to fall. At any given moment, some 2,000 thunderstorms roil the atmosphere around the globe. A thunderstorm lasting several hours generates energy equivalent to that of a small hydrogen bomb, whereas a hurricane releases an equivalent amount of energy every second.

As the sun heats the land and oceans, surface water evaporates and rises as water vapor into the troposphere, sometimes to tremendous heights. There it cools, condenses, forms clouds, and eventually falls back to the planet's surface.

Near the Equator, enormous masses of air are heated, causing them to expand and carry moisture-laden air upward and poleward.

Dry air masses moving toward the Equator from the polar regions are deflected by the Coriolis force, which is an effect created by the earth's rotation. When the air masses of different temperatures and humidities collide, they generate the atmospheric turbulence that causes our changing weather. The planet's rotation also affects the winds circulating between the Poles and tropical regions, the high-altitude meandering jet streams, and the trade winds blowing toward the Equator.

Swiftly moving jet streams loop around the planet like high-flying ribbons of air some five to ten miles above the earth's surface. On a typical day, for example, the jet stream over North America undulates eastward, marking the approximate boundary between cold, dry air masses and warm, humid ones. It provides steering currents for low-pressure cyclones that travel along the frontal boundaries between the two air masses of higher pressure. Pushed eastward by prevailing westerlies, these air-mass fronts and low-pressure systems move across the continent, bringing alternating periods of fair weather and storms.

In the Northern Hemisphere, winds move clockwise around high-pressure systems and counterclockwise around low-pressure systems; in the Southern Hemisphere winds spin in the opposite directions. Over equatorial ocean waters, moisture-rich trade winds of one hemisphere collide with those of the other in the Intertropical Convergence Zone, spawning wide bands of cumulus clouds and tropical storms such as hurricanes. In places where air masses of different temperatures and humidities merge, a "front" forms and causes stormy weather. When cold, dry air displaces warm, moist air, the cold front normally dominates; a warm front occurs when warm, moist air overrides cold, dry air.

Storms forming over the oceans generally dissipate over time or become organized into systems. Such storm systems either die at sea or, if tropical in origin, find themselves fueled by warm, rising ocean mists that add heat and moisture. Such systems can evolve into independent tropical storms called hurricanes, which have sustained winds greater than 74 miles an hour until they move over land and their warm-water fuel supply is choked off.

As part of the weather machine, winds provide an extremely efficient mechanism for redistributing heat more evenly around the planet. The sun warms the equatorial regions more than the polar ones, where long periods of darkness prevail during the winter months. The excess heat in the tropics is borne away by upper-level winds that carry it toward the higher latitudes. When the warmer air reaches the middle latitudes, however, it begins to cool and lose its tropical characteristics. Polar air moving into the middle latitudes is warmed and eventually loses its polar identity.

Because the earth is a wet planet, clouds are often present even on seemingly clear days. Composed of small ice crystals or suspended water droplets, clouds come in a variety of sizes, shapes, and colors. In 1803, British scientist Luke Howard first classified clouds into three categories: cirrus, cumulus, and stratus.

Curly, wispy, and stringy, high-flying cirrus clouds—floating some four miles above sea level and composed of ice crystals—are often precursors of a distant approaching storm. Puffy, white cumulus clouds signal fair weather, in contrast to the lumpy and gray cumulus clouds that often precede storms. Cumulus clouds form when updrafts of warm, moist air slowly cool a little less than a mile above the planet's surface. At first puffy and white, they can build into the charcoal-colored cumulonimbus clouds that characterize thunderstorms, producing lightning, hail, and torrents of rain. More stable stratus clouds form at the boundary of a warm front and appear when warm saturated air overruns cold air masses. Stratus clouds often form a gray layer and gently blanket the sky for hundreds of miles, especially in the polar regions. The precipitation that falls from them is usually in the form of drizzle or light snow.

RICHARD LEWIS/WEATHERSTOCK

Eventually water falls from the atmosphere, and much of it ends up in the oceans directly or by runoff from rivers. The precipitation that falls on land, while occasionally hazardous,

In 1988, a dark cloud hovers ominously over Cleveland, Ohio. Called a roll cloud because it can roll slowly about its horizontal axis, this relatively rare cloud forms where cool air moves gently into drier, warmer air. Such sites include gust fronts and the rear of storms. Completely detached from the base of a thunderstorm, a roll cloud assumes a long, narrow, tube shape.

also sustains life and plays a major role in determining what species survive.

Land receives water unevenly from the sky. At Mawsynram, in Assam, India—considered to be the earth's wettest location—the annual average rainfall is 467 inches, in contrast to Chile's Atacama Desert, where the parched town of Arica enjoys a barely detectable rainfall amount measuring only two-hundredths of an inch per annum.

WEATHER / WILD WINDS

The wind plays a major role in shaping the earth's weather. Because the sun heats our rotating and tilted planet unevenly, vast regions of high and low pressure form as great masses of air warm, expand, and rise while others cool, contract, and sink. Wind performs the function of a great equalizer, as large amounts of air migrate from high- to low-pressure zones, sometimes taking the form of gentle breezes and other times becoming destructive gales.

In the Northern and Southern Hemispheres, three kinds of large-scale wind patterns dominate atmospheric conditions and affect the weather: the polar easterlies, the temperate westerlies, and the tropical trade winds. As warm equatorial air rises and migrates toward the Poles, cooler air moves in to take its place. When warm equatorial air reaches roughly 30° latitude, it begins to cool and sink, with some curling back toward the Equator while the rest moves poleward. At about 60° latitude, polar air collides with and lifts warmer air moving from the Equator.

Winds tend to blow west and east, rather than north and south, because the earth's eastward rotation and the subsequent Coriolis effect cause winds to curve to the right north of the Equator and to the left in the Southern Hemisphere. When Northern Hemisphere air, for instance, travels south toward the tropics, it gets twisted to the west, thus generating northeasterly trade winds that once ferried sailing vessels from Europe to North America. In the zone where the hemispheric trade winds collide, they tend to weaken, becalming wind and water into doldrums that affect sailing vessels.

The seasonal oscillation of the area where trade winds meet—also called the Intertropical Convergence Zone (ITCZ)—helps bring on conditions for periodic torrential rains. When an atmospheric disturbance, or easterly wave, forms on the ITCZ, downpours result, causing severe flooding over islands. These waves can evolve into tropical cyclones. Seasonal monsoon rains, which bring moisture to farmlands, can drench India, Myanmar, and Bangladesh. In one extreme, between August 1860 and July 1861, monsoon rains soaked the town of Cherrapunji in the foothills of the Himalaya with a record 1,042 inches of rainfall.

While large-scale wind patterns shape and influence the weather across entire continents, other winds limit their effects to regional and local areas. Such breezes can sweep down mountains, rush through valleys, whip across deserts, or blow past shorelines. The warm, dry foehn, for example, hugs mountain slopes in central and northern Europe. The North American chinook, also warm and dry, descends the Rocky Mountains' eastern slopes and blows across the western plains, quickly raising the temperature.

Some local winds can be particularly brutal. In winter, the buran brings blizzards to Central Asia and Siberia. France's Mediterranean coast is lashed by the mistral most often during winter and spring. In Texas the norther, or *El Norte*, as it is known in Mexico and Central America, roars in and chills the air within hours. In Uruguay and Argentina, the cold, dry pampero sweeps over the grasslands, or pampas.

In springtime, the sirocco blows out of the Sahara, crosses the Mediterranean Sea, and

On February 17, 1988, fierce Santa Ana winds gusting to 80 miles an hour blew across southern California. North of Los Angeles, the strong winds tipped a tractor trailer against the guardrail on the Foothill Freeway. Santa Anas can last a day, a week, or longer.

strikes peninsulas in southern Europe with very warm, humid, and dusty winds. The hot brick-fielder blasts out of central Australia and lofts tons of dry soil from brickfields near Sydney.

Over the years, the Santa Ana winds blowing across southern California from the mountains and desert have fanned small brushfires into fierce firestorms. The hot winds eventually dessicate everything in their path, transforming landscapes into fodder for wildfires. Tragically, expensive beach homes and coastal estates have been turned into charcoal monuments to incendiary Santa Ana breezes.

In October 1991, flames engulfed 1,500 acres of northern California hills in the cities of Oakland and Berkeley, near San Francisco. Warm, dry gusts from the north fanned advancing fires, driving them southeastward through residential neighborhoods. As houses burst into flames, their burning cedar shingles flew like Frisbees across streets and highways and ignited neighboring homes. Ultimately fire ravaged some 3,000 homes and businesses and killed 25

people, a high toll, especially in the wake of the Bay Area's 1989 Loma Prieta earthquake.

Less than a year later, during August 1992, a conflagration in central California's Calaveras County claimed 17,000 acres and destroyed 170 homes and buildings. During the summer of 1994, the Wenatchee National Forest and other parts of Chelan County, in the central region of Washington State, lost 186,694 acres to blazes, when winds whipping through steep canyons lofted flames over tree-covered slopes and across highways, thwarting firefighters who struggled to contain them.

Forests dried by steady breezes of warm air are especially vulnerable to lightning strikes. Like tinder, dead trees will quickly ignite when struck by nature's fire. In 1988, for example, Yellowstone National Park, which extends from Wyoming into Montana and Idaho, suffered devastating forest fires after a summer of hot winds and drought turned dead lodgepole pines into ready kindling. When lightning strokes sparked natural forest fires, park officials found themselves unable to control the flames, which spread wildly and spilled into neighboring territory. In two parks and nearby forests, 1.4 million acres burned and 60 homes and historic buildings suffered fire damage.

CLIMATE

Earth is a planet of extremes, with heat and heavy rain in its tropics; subzero temperatures, winds, and ice in its polar regions; and an extraordinary spectrum covering a full range of alternatives in between. San Francisco and Beijing, for example, lie in the midlatitudes. Both places record similar annual temperature and precipitation ranges, but their seasons are quite dissimilar. The Bay City experiences dry summers and wet winters, while the Chinese capital has wet summers and dry winters. Winter in San Francisco is not much cooler than summer. Beijing, on the other hand, is hot in summer and cold in winter.

A region's average weather conditions over a long period of time, say 30 or so years, define its current climate. In its nearly 4.7-billion-year history, the earth has undergone tremendous climate variations, and although it now enjoys a relatively warm spell, the existence of ice sheets in Antarctica and Greenland reminds us that the planet remains in a partial ice age.

Geological evidence reveals that the overall climate and those of particular oceans and continents have changed drastically and repeatedly during the past few million years. Ridges of glacier-deposited rock and debris, called moraines, show that ice sheets covered parts of the Northern and Southern Hemispheres at least ten times during the past one million years. Fossil shells of extinct marine animals have turned up in the Sahara, indicating that an ocean covered today's desert some 400 million years ago. Leaf fossils resembling modern-day tropical plants found scattered in sites around the globe suggest that similar plants once flourished worldwide.

Today, life is still adapting to climate variations. The conflict between changing climates and an organism's ability to adapt has spawned an extraordinary diversity of life-forms. Climate influences everything from food choices to mating patterns to the traits necessary to survive. People, no less than plants and animals, must adapt to climatic variability, building structures appropriate for warmth or cooling, planting crops in harmony with weather patterns, and creating cultures that relate to natural cycles.

In 1900, Wladimir Köppen, a Russian-born meteorologist working mostly in Germany, devised a classification system for climatic regions. Using temperature and precipitation data, Köppen proposed six climate groups: tropical, dry, mild, continental, polar, and high elevation. Those groups were subdivided into 14 climate types, including wet and dry, arid, semiarid, tundra, and ice cap.

In places with predictable wet climates, such as Hawaii and equatorial West Africa, rain falls nearly every afternoon, with annual rainfall of some 60 inches watering lands whose average monthly temperatures rarely vary from the expected 77°F to 82°F. Such climates reach some 10° latitude north and south of the Equator, a band continually affected by the moisture-rich Intertropical Convergence Zone (ITCZ), where the trade winds of the Northern and Southern Hemispheres meet.

The ITCZ periodically swings north and south of the Equator, showering regions outside that band with rain only part of the year. These tropical wet and dry climates enjoy three seasons—cool and dry, hot and dry, and hot and

PETER-NOEL WEBB/OHIO STATE UNIVERSITY

wet. Because life here depends on rain, dry years can cause great hardship. The cities of Calcutta, India, and Havana, Cuba, and Africa's Serengeti Plain are in this climate zone.

In arid and semiarid climates, precipitation is low. Rainfall totals only 4 to 12 inches a year in arid regions and 12 to 24 inches in semiarid ones. And hot days can be capped by cool nights. Arica, a town in the arid coastal region of Chile's Atacama Desert, one of the earth's driest spots, receives two-hundredths of an inch of yearly rainfall. In Libya, the town of Azizia often heats up, reaching 136°F on September 13, 1922, an all-time weather record. Because of their positions relative to ITCZ swings and midlatitude low-pressure systems, Africa's Sahara and Australia's outback—in arid regions of the subtropics—rarely see rain clouds form. In the U.S. and other lands where mid-continent mountains block ocean moisture, regions such as the Great Plains can have dry climates.

In the earth's temperate regions, continental climates enjoyed by Canada and the northern U.S., for example, bring changing seasons marked by cool summers, snowy winters, and short growing periods—the result of ongoing battles between tropical and polar air masses. Climate is also influenced by latitude and position on a continent. While cities such as Los Angeles and Jerusalem bask in the warm, dry summers and the brief, rainy winters of a Mediterranean climate, cities such as Savannah, Georgia, and Shanghai, China, swelter in hot and humid summers and shiver in winter cold spells common to humid subtropical climates. Such variations contrast with the gray drizzle of Seattle, Washington, and Wellington, New Zealand, which lie in west-coast regions having milder maritime climates.

Only in the Northern Hemisphere do three types of continental climates occur. They range from warm summer to cool summer to subarctic. Catching south-sweeping Arctic winds in winter, these regions, including New England in the U.S., and parts of northern Europe, can suffer severe cold and blizzards, such as the one that felled Napoleon's troops fleeing Russia in 1812. With the drama of bitter winters, powerful thunderstorms, and tornadoes, however, also comes beauty. People who live in these climates can rejoice in some of the planet's most spectacular fall colors.

The two polar climates, tundra and ice cap, mark the earth's coldest places. While the Arctic and Antarctic regions remain frozen most of the year, tundra regions can host brief summers, enabling birds to gorge on insects from fields dotted by wildflowers. Frigid winds and glaciers can also chill regions near the Equator, in the highland climate of mountains such as Africa's Kilimanjaro or South America's Andes.

CLIMATE / THE MEDITERRANEAN

Given the Mediterranean Sea's remarkable role in human history, it is no surprise that this "sea in the middle of land," as its Latin name attests, should also have earned its status as "the incubator of Western Civilization." Occupying a deep long depression, the sea reaches 2,300 miles from the Strait of Gibraltar in the west to the Gulf of Iskenderun in the east, and measures, on average, 500 miles from north to south.

The Mediterranean's historical drama hardly compares with its geological drama, however. Here, an area became a sea, evaporated into a desert, then flooded again. And each metamorphosis altered the region's climate, affecting annual rainfall amounts in Europe and Africa. Today, shallow, narrow channels link the sea to the Atlantic Ocean, the Black Sea, and the Red Sea. But this geography tells only a relatively recent geological tale.

Some 200 million years ago, when the great Pangaea landmass fractured into separate continental plates, the ones carrying Africa and Eurasia pulled apart and a large body of water, part of the Tethys Ocean, occupied the area.

Then, about six million years ago, the African plate moved northward and collided with Europe, plugging the channel between the Atlantic and Tethys Oceans and choking off the water supply from the west. When the plates converged, the Tethys shrank, leaving a remnant

The eight-mile-wide Strait of Gibraltar, visible in this image from space, separates Europe and Africa. Tidal pulses of water from the Atlantic Ocean, at lower left, enter the strait and trigger wave clusters that expand eastward across the Mediterranean Sea.

that became the eastern Mediterranean and Black Seas. Plate movement also carved a chasm for a new sea in the western part, and when a channel opened at Morocco, seawater poured though it and filled the new depression.

The leading edge of the African plate plunged under the Eurasian plate, and the collision separated parts of North Africa, created the Alps of Europe, and split the islands of Corsica and Sardinia from France and Spain. Volcanoes formed along plate boundaries near Sicily. This collision continues; volcanoes still erupt and earthquakes occasionally shake the region.

Along the eastern Mediterranean Ridge—a raised scar resulting from the African plate's subduction—the Aegean Sea now covers Asia Minor's drowned tip. As the African plate tugs on the Aegean seafloor, the island of Crete, riding the Eurasian plate's leading edge, is being pulled toward Libya. Their eventual collision will create a new range of mountains in Africa.

While cutting off the Tethys water supply, the collision coincided with other geological events that caused the sea to evaporate and turn the deep Mediterranean basins into regions of vast salt deposition. "The sheer enormity of this event, known as the Messinian Salinity Crisis, is without modern analogy," says Richard H. Benson, a senior scientist at the National Museum of Natural History. "Marine ecosystems were destroyed and layers of salt and gypsum—as much as a mile thick in some places—were deposited on the bottom. Giant supersaline lakes, fed by the rivers of Europe and Africa, came to occupy the deep Mediterranean basins and new species of lake-dwelling creatures came to inhabit these shallow waters."

Scientists discovered the magnitude of this phenomenon two decades ago by drilling into seafloor sediments to obtain long core samples, collecting rocks and fossils, and using seismic and sonic probes on the deep seabed. Because of the salt bed's thickness, Benson and others believe that Atlantic water continued to flow into the Tethys despite partial channel blockages, creating a salt trap. So thick are the salt deposits that the Mediterranean may have filled and evaporated cyclically many times over 450,000 years, accumulating 6 percent of all ocean salt, Benson speculates.

Struggling to explain deep-sea and planktonic microfossils found in sediments laid down before and after the great salt deposits, Benson's research team first happened upon the Messinian Salinity Crisis. Puzzled by fossils of shallow lake-dwellers in late Messinian sediments, plus an absence of marine fossils in the youngest Messinian rock, the scientists pieced together a series of events to match the vanishing of the Tethys Sea.

Ultimately, the barrier blocking the Atlantic Ocean from the Tethys's isolated basins broke, resulting in rapid flooding known as the Zanclean deluge, which filled the modern Mediterranean. The event that prompted the Atlantic to pour water into the Mediterranean basin remains unknown, because no firm geologic record has yet been uncovered. "Explanations of how these events took place are still forming," says Benson. "Its importance to our understanding of the natural events that are unfolding today cannot be overestimated."

To help pin down more precisely the dates of these geologic occurrences, paleobiologists

Six million years ago, during the Upper Miocene epoch, the African plate plunged under the Eurasian plate, temporarily stopping the flow of Atlantic water into the Mediterranean Sea and transforming the seabed into a great salt flat. In this artist's rendition, four-tusked elephant ancestors, or **Gomphotheres,** *migrate over the dried-up basin toward Europe, where a new volcano rises atop the overlying Eurasian plate.*

ROBERT E. HYNES

have turned to other specialists for assistance. Using information from paleomagnetic studies of the shifting North and South Poles—tied to rhythms in the earth's orbital motion—the scientists have pegged the start of the Messinian Salinity Crisis at 5.8 million years ago, a figure varying from previous geologic records by 400,000 years. By using the same astronomic stratigraphy, a dating method that studies perturbations in the earth's orbit from cyclic sediments in the ocean, a team of geologists from Holland set the start of the Pliocene epoch, which coincided with the Mediterranean's flooding, at 5.35 million years ago.

GREAT DISASTERS

Land and sea may lie relatively still for centuries. Then, tragically, a geologic upheaval suddenly unleashes a cataclysm. Perhaps no event illustrates such horror so clearly as the calamity at ancient Thera. What happened there may have been the most explosive volcanic eruption in recorded history.

Towering over the Aegean Sea some 3,500 years ago, the mile-high volcano formed a ten-mile-wide island. Thera provided a base for the magnificent Minoan civilization centered 70 miles south on Crete. At its peak, 30,000 people dwelled in Akrotiri, Thera's main city, erecting frescoed art-filled palaces and sending out ships laden with trade goods.

While historians remain uncertain of the exact date—estimates range from 1470 to 1628 B.C.—they know the sequence of events. Light earth tremors were followed by a violent quake, aftershocks, and an explosion that blew the top off the mountain. Scientists think that the booms were audible as far as Scandinavia, the Persian Gulf, and the Rock of Gibraltar.

Thera's cone collapsed in glowing avalanches, unearthing the magma chamber, which quickly filled with seawater. Then an immense explosion eradicated 32 cubic miles of the island and left a caldera of 30 square miles. So forceful was this blast that tsunamis 160 to 300 feet high struck Crete and caused terrible destruction. Some scholars say the event may have led to the 1450 B.C. collapse of the Minoan civilization, thus ending 15 centuries of dominance in the region.

Also included among history's deadliest natural events is the Lisbon earthquake of November 1, 1755. At roughly 9:40 a.m., ominous sounds "resembling the hollow distant rumbling of thunder" heralded 3 major shocks in 15 minutes. Seismic waves originating on the floor of the Atlantic Ocean focused their energies on the city, where 17,000 houses crumbled. Chandeliers swung wildly in Lisbon's cathedral. There and at the Church of St. Anthony, parts of the buildings collapsed, entombing hundreds of worshipers crowded into the cathedral square. When the third tremor struck, Portuguese eyewitness Antonio Pereira said, "The whole tract of country about Lisbon was seen to heave like the swelling of the billows in a storm."

Within minutes tsunamis estimated at 40 to 50 feet high arched up like roiling muddy mountains, roared up the Tagus River, and smashed the waterfront, sucking screaming victims to their deaths. Then came the fires, which consumed Lisbon in a firestorm that lasted five or six days. The opera house, Patriarchal Church, and Royal Palace—including a 70,000-volume library—were consumed by fire.

While earthquakes and volcanoes have secured places in history as indiscriminate claimers of life, floods have taken more lives than any other form of ordinary natural disaster. And no country has lost more lives to flooding than China, primarily along the banks of its principal rivers, the Huang Ho and Yangtze.

China's worst flood occurred in September and October 1887, when a dike at a sharp bend in the Huang Ho collapsed after weeks of heavy rain. The break swelled to hundreds of feet, then to half a mile as dike walls eroded away. Yellow torrents swallowed whole towns in Henan and Shandong provinces, burying homes in silt and sweeping away 1,500 communities. Thousands of residents drowned, while dazed victims

Whitewashed buildings cling to Thera's cliffs, where residents enjoy spectacular views of the Aegean Sea. Once part of an ancient volcanic island, Thera and nearby isles acquired their crescent shapes after a massive central volcano exploded some 35 centuries ago. The volcano collapsed in on itself, forging a caldera that became the foundation for today's craggy islands and their deep harbor.

GORDON GAHAN

clutched stationary objects. Reportedly, more than 900,000 people died—some estimates surpass 2.5 million—while more than 2 million people became homeless.

Fortunately, an early warning system that uses satellites to see approaching storms has helped reduce suffering in countries vulnerable to floods, and except for war, death on a grand scale seems unimaginable today. But the possibility of a devastating natural disaster always lurks. A strike from an errant large asteroid, for instance, could kill millions of people.

GREAT DISASTERS / IMPACTS

Just off the coast of Mexico's Yucatan Peninsula, near the village of Chicxulub, an extraordinary impact took place 65 million years ago. Many scientists believe that because of the large number of terrestrial and marine species that became extinct, it may have played a crucial role in the subsequent evolution of life on the planet.

Hurtling through the atmosphere from space, a comet or an asteroid six miles in diameter smashed into coastal waters near the Yucatan's flatlands, blasting open a crater and rocketing billions of tons of rock, dust, and seawater vapor into the upper atmosphere. So powerful was the blast that it quickly heated the atmosphere, ignited forests, and sent tsunamis crashing onto coasts. A veil of dense dust likely plunged the earth into months of darkness; acid rain may have killed most of the surviving plants and animals.

In 1980, father-and-son scientists Luis and Walter Alvarez and two colleagues proposed that an asteroid six miles in diameter had struck the earth. At sites around the world, they found iridium concentrations in rock layers that all dated to the same time. Iridium exists as mere traces in the earth's crust but is more abundant in extraterrestrial rocks. Such an impact, scientists argue, accounts for the disappearance 65 million years ago of two-thirds of the planet's species, including all dinosaurs.

Because asteroids, smaller rocky fragments, and comets are common in the solar system, it is no surprise that craters pock the faces of planets and moons. During the nights of annual meteor showers, earth-bound sky gazers may easily see a meteor every few minutes. Asteroids, which

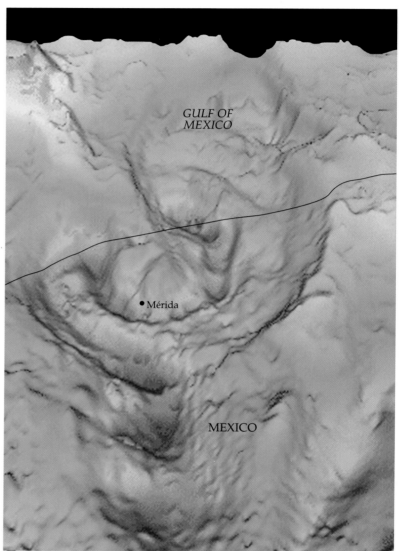

GULF OF MEXICO

●Mérida

MEXICO

can span hundreds of miles, orbit mostly within the solar system's main asteroid belt, but occasionally two collide and their fragments adopt new orbits, some of which may cross that of the earth. Comets, giant clusters of rock and ice, sometimes cross earth's path.

A computer-generated gravity-field map of the Chicxulub crater (opposite) points to a cataclysmic impact in Mexico's Yucatan Peninsula. The crater's large size and iridium found in rocks around the world support the theory that an asteroid struck the earth some 65 million years ago. The impact (below) triggered a series of catastrophic events that might have caused the extinction of more than half the planet's species.

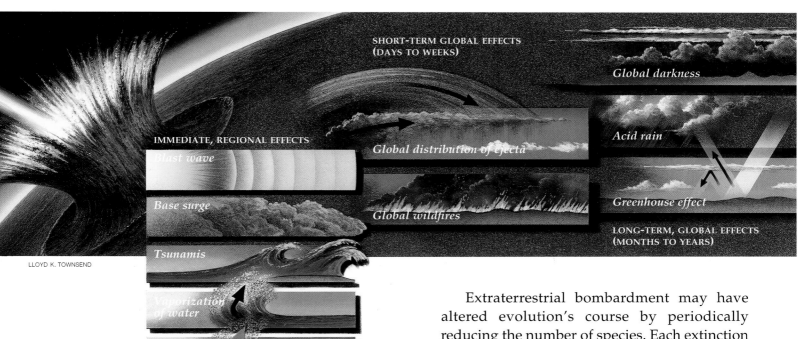

SHORT-TERM GLOBAL EFFECTS
(DAYS TO WEEKS)

Global darkness

Acid rain

IMMEDIATE, REGIONAL EFFECTS

Blast wave

Base surge

Global distribution of ejecta

Greenhouse effect

Tsunamis

Global wildfires

LONG-TERM, GLOBAL EFFECTS
(MONTHS TO YEARS)

Vaporization of water

Vaporization of rock

Earthquakes

LLOYD K. TOWNSEND

At some point during the next million or so years, the earth may get hit by a large asteroid. The planet has undergone repeated impacts since forming nearly 4.7 million years ago, but wind, rain, and crust remodeling have buried or eroded most craters. Scientists only agreed in the 1960s that a 150-foot-wide asteroid created Arizona's Barringer Crater (or Meteor Crater) some 50,000 years ago. They based this finding on geologist Eugene Shoemaker's evaluation of the site's shattered rock, iron fragments, and melted glass. Researchers have since documented more than 150 impact structures worldwide.

Extraterrestrial bombardment may have altered evolution's course by periodically reducing the number of species. Each extinction event, however, has been followed by radiation of surviving species into a plethora of new ones. Dinosaurs, for example, ruled the earth for 140 million years before their extinction.

Earthlings got edgy in May 1996, when a quarter-mile-wide asteroid zipped by only 280,000 miles away—nearly the distance from the earth to the moon. Most impressive, too, were the multiple plunges into Jupiter's atmosphere by fragments of comet Shoemaker-Levy 9 in July 1994.

Fortunately, most asteroids smaller than 150 inches across break up in the atmosphere. Asteroids measuring 300 to 3,300 feet across, however, strike about every 1,000 to 200,000 years. When one hits land, it will create a massive crater; if it hits the sea, it could drown millions of people in tsunamis. About every 300,000 years a mile-wide asteroid is likely to impact the earth.

LIVING IN HARM'S WAY

Perched on one of the world's largest river deltas, where the Ganges, Jamuna, Padma, and Meghna Rivers merge, and poised at the head of the Bay of Bengal, Bangladesh is visited yearly by floods, severe storms, and storm surges. During the past 40 years, nearly 20 cyclones have struck the country. Among them were 7 of the century's 9 deadliest storms, which collectively claimed more than half a million lives.

On April 29-30, 1991, a cyclone's 145-mile-per-hour winds and 20-foot storm surge swept the heavily populated delta and floodplain between Chittagong and Cox's Bazaar, killing 139,000 residents and leaving 10 million people homeless. The tragedy recalled 1970, when a cyclone sweeping across the same low-lying areas killed 300,000 Bangladeshis. Facing a history like this, one must ask: Why live there? Why dwell in harm's way?

Among the earth's most densely populated countries, Bangladesh packs some 1,500 people into every square mile—a density equivalent to placing half the population of the United States into Wisconsin. Lacking sufficient arable land, many of the country's 110 million people cram themselves onto islands and fragile "chars"—mounds of silt that may last from months to years before washing away. Though transient, the chars provide fertile farmlands for growing rice and vegetables. And after every catastrophe, residents return to farm them again.

Some wealthier westerners scoff at this behavior, but many of them indulge in a similar catastrophe roulette. Consider Mississippi riverbank dwellers who, during the summer of 1993, found themselves swimming through their homes. By summer's end, 1,100 levees had failed and an estimated 50 people had died.

In recent years, floodwaters have also brought suffering to Europe. On September 22, 1992, a 49-foot-high wall of water from the swollen Ouvèze River plowed through southern France's Vaison-la-Romaine, drowning 38 people. During Christmas 1993, rain and snowmelt flooded western Europe. Then, in January 1995, the Meuse River submerged Redon, France, and sent water pouring into the Netherlands.

In South Florida, hurricanes have killed thousands of people since the early part of this century. Florida's turmoil matches that of southern Texas, also vulnerable to hurricanes, tornadoes, and storm surges. In the midwestern U.S., some 300 tornadoes each year take lives and property in Tornado Alley. And in northern states, winter's blizzards besiege cities and farms.

Consider the California cities of San Francisco and Los Angeles, where wealthy residents build expensive homes perilously close to the San Andreas Fault. Movement along the 750-mile-long fault, which has killed thousands of people over the years, continues to jeopardize millions of homes and lives. Scientists have calculated quake probabilities for its various segments, and although the time brackets are wide, they are helpful to planners who must prepare for the inevitable "big one."

The San Andreas Fault is not the only problem. A complex network of branching faults extends from the San Andreas system, as residents have discovered recently. On January 17, 1994, the moment-magnitude 6.7 Northridge earthquake killed 60 people, destroyed more than 3,000 homes, toppled 10 bridges, and

Hurricane Elena blew in from the Gulf of Mexico on Labor Day weekend, 1985. In Alabama, the powerful storm damaged about 80 percent of Dauphin Island's 800 homes, sweeping away 50 of them. Elena also struck Mississippi, Louisiana, and Florida with 125-mile-per-hour winds. In the end, total insured losses neared $543 million. The quick evacuation of 1.5 million residents helped to minimize fatalities from the storm.

made 3 freeways impassable. Scientists say higher-magnitude quakes likely will hit Los Angeles during the next 30 years.

Japanese citizens, no less than Californians, face geologic hazards every day, because many of their homes lie near or at the foot of volcanic slopes in some of the earth's most volatile earthquake zones. A person has to think only of the city of Kobe to appreciate deadly movement along a minor fault. There, at 5:46 a.m. on January 17, 1995, a violent quake shook Japan's sixth largest city for 20 seconds, toppling elevated highways, collapsing homes and businesses, and twisting railways while killing 5,200 people, injuring 25,000 more, and leaving 300,000 residents homeless. The event was a painful reminder of the country's Great Kanto Earthquake of 1923, which claimed 143,000 lives—30,000 of whom died when a firestorm engulfed a park packed with people.

In July 1993, an undersea earthquake off Hokkaido, Japan, gave residents only minutes to escape a tsunami that surged onto the island of Okushiri, killing more than 200 people, crushing houses, and tossing boats ashore. In Aonae, half of the town's 680 homes were leveled when firestorms from ruptured gas lines burned them to the ground.

Japan's complex geology makes it vulnerable to volcanic eruptions as well as earthquakes. In 1991, for example, Mount Unzen began erupting, spilling red-hot pyroclastic flows down the mountainside. One flow killed 43 people. At the foot of Sakurajima, residential communities occasionally find themselves bombarded with tephra. At Shimabara, heavy rains have crushed and smothered nearby subdivisions in rivers of mud.

Consider residents living near Hawaii's Kilauea. The erupting volcano has added 500 acres of new land since 1982. While increasing the island's size, Kilauea's lava flows have consumed the community of Kalapana, of which little remains but lava remnants. Such is the price for settling atop one of the earth's most active hot spots, which spawned the long chain of Hawaiian Islands and built it up over millions of years.

In Washington State, homeowners often cast anxious eyes toward the Cascade Range. Some 5,000 years ago, landslides from Mount Rainier, a dormant volcano, spread mudflows over the Puget Sound lowlands, including part of the area where the present city of Tacoma sits. More than a dozen potentially hazardous nearby volcanoes remain active, including Mount Baker and Mount St. Helens, whose 1980 blast killed 57 people and flattened trees for 19 miles.

Innocent-looking peaks can suddenly become active volcanoes after hundreds of years of quiet, quickly disrupting human activity. On June 15, 1991, Mount Pinatubo in the Philippines exploded, shattering a thousand feet of rock and blowing clouds of tephra, water vapor, and aerosols of sulfuric acid ten miles into the sky. Nearly a thousand people died during the eruption or its aftermath, and the toll would have been much higher if scientists had not predicted the eruption, allowing thousands to be safely evacuated. Yet the people who live nearby still suffer from lahars—floods of mud that can cover fields, roads, and houses.

Ecuador's majestic Cotopaxi, one of the world's highest active volcanoes, is currently quiet, but scientists say that heat released in a small eruption could melt enough glacier ice and snow to pour torrents of mud, or lahars, into lowlands below. In November 1985, Colombia's Nevado del Ruiz produced a minor eruption, sending a 130-foot-high wall of mud through the Lagunilla River's canyons and burying 20,000 sleeping residents of Armero, an agricultural center. Another nightmare was

born on May 31, 1970, when an earthquake in Peru shattered Nevado Huascaran's western face, a peak 8 miles from the town of Yungay. Within 3 minutes, a debris avalanche had buried about 18,000 of Yungay's residents in nearly 30 feet of rock.

The Italian village of Zafferana, on Sicily, lies in the ominous shadow of Mount Etna, which awoke in 1991 and poured glowing lava down its slopes for 473 days. Trying to save Zafferana, Italian troops detonated explosives on the mountainside to divert the lava flow from its natural channel. U.S. Marines helped out by airlifting concrete dams that were put in place to block the flow along the channel. Finally, it was diverted into an excavated channel, and the natural one was blocked.

Almost any area near an ocean can receive a seemingly random wall of water called a tsunami, which can be triggered by an undersea earthquake, eruption, or landslide. According to the National Oceanic and Atmospheric Administration, tsunamis killed more than 2,000 people along the coasts of Nicaragua, Indonesia, and Japan in 1992 and 1993 alone. In 1960, a Chilean quake sent a tsunami across the Pacific Ocean, destroying life and property from Chile to Hawaii and Japan. One of history's worst disasters occurred in 1883 when Krakatau, in the East Indies, erupted explosively, generating a sound wave heard 3,000 miles away and creating tsunamis up to 125 feet high. The waves razed 165 coastal villages and killed 36,000 people.

Why dwell in precarious lands and live with the possibility of disaster? Everyone has reasons—natural resources, business, farming,

Facing a restless giant across a narrow bay, residents of Kagoshima, a thriving port city in Japan, find themselves regularly dusted by ash from Sakurajima, one of the world's most active volcanoes. The 7,500 people who live near the volcano's base prepare for eruptions by conducting annual drills. Local students frequently must wear hard hats to protect themselves from falling debris.

ROGER RESSMEYER / CORBIS

land ownership, or family history, to name just a few. Some stalwarts remain there because their families have done so for generations and they wish to preserve tradition.

There is yet another, more elusive, motivation: Some of the earth's most dangerous spots are extraordinarily beautiful. People have long sought and settled these stunning settings, taking great pride in their glorious homelands. Moreover, no location is completely without risk. Every place presents some kind of hazard and eventually exacts a price. Such is the nature of life itself, and in some sense, to live without risk is to not live.

COPING WITH CATASTROPHE

Human beings are resilient, sometimes astonishingly so. When an unexpected natural catastrophe occurs, survivors somehow pick up the pieces and move on.

Residents of high-risk regions accept the fact that they have chosen to live with risk, in some cases suppressing a background of anxiety and fear. Many Californians, for example, knowingly await "the big one," an earthquake measuring 8.2 or larger, which geologists say will strike the Los Angeles region during the next few years. The trouble is, many scientists

When the Coeur d'Alene, a river in northern Idaho, surged eight feet above flood stage in February 1996, many residents found their homes and businesses damaged or destroyed. Below, Ed Conley, an official with the Federal Emergency Management Agency, talks with two boys as he surveys the scene and tells families about assistance programs.

expect that it will come soon, but no one is quite sure what "soon" means.

Where the North American and Pacific plates converge in California, strain along the San Andreas Fault system steadily builds and then eases in sudden slippages—like a brick dragged by a spring along a carpet. The metaphoric spring-force builds, and eventually the brick leaps forward, only to wait again until the steadily mounting force overcomes the friction holding it in place. This "strike-slip" motion characterizes many faults, leaving residents who live near them waiting anxiously in the lulls between sudden earth movements.

In January 1994, just after the moment-magnitude 6.7 quake at Northridge, California, many survivors chose to camp out rather than sleep in their damaged houses. Aftershocks kept them on edge for weeks. Although northern California's Loma Prieta quake, in October 1989, wielded a moment-magnitude of 7, the larger quake caused less damage. The reason for the difference lies in location. The quake at Northridge, near Los Angeles, focused its energy under a heavily populated commercial zone, built on weak sedimentary rock that shook more violently than the denser, firmer rock at the Loma Prieta epicenter.

The Loma Prieta quake damaged about 1,000 buildings and collapsed a major freeway. The Northridge quake damaged or destroyed more than 3,000 homes, 10 highway bridges, and 3 freeways. Comparing these quakes with a coming "big one" leaves Californians disturbed and wondering what to do.

Another sample of what may result from movement along the Pacific and North American plate boundary was seen in Alaska on March 27, 1964. The Good Friday quake, with a 9.2 moment-magnitude—among the largest ever recorded—fractured the Pacific plate. Faulting raised the seafloor 38 feet and sent a tsunami toward Anchorage. The earthquake and tsunami claimed 131 lives and resulted in property losses of 750 million dollars.

Understanding that a dollar of prevention can truly prove to be worth a thousand or a million dollars in cure, people in high-risk locations are increasingly interested in early detection and warning systems.

To spot tsunamis before they surge inland, authorities established the Pacific Tsunami Warning Center after a 1946 Alaskan earthquake produced sea waves that killed 159 people and caused major damage in Hilo, Hawaii. Monitoring stations scattered around the Pacific Rim send seismic and tide-level information to the center, operated by the National Oceanic and Atmospheric Administration (NOAA). It can warn coastal residents an hour before a tsunami strikes land.

In Alaska, a second warning center sends regional alerts to coastal dwellers from Alaska to southern California. Japan, Russia, French Polynesia, and Chile have also established warning systems. Currently, NOAA is testing an automated seismic warning system. When vibrations of sufficient magnitude are detected, a computer signals a satellite that contacts stations in affected regions, and alarms then sound. Warning time is cut to two minutes after the tsunami has been generated.

Geologists estimate that some 500,000 detectable earthquakes take place globally each year, but humans can sense only about 100,000. Of these quakes, approximately 1,000 are capable of damaging property. While some earthquakes have killed thousands at a time, a few of them have taken hundreds of thousands of lives. A large earthquake shook China in 1556, for example, and killed an estimated 830,000 people. Another one rattled the region around the eastern Mediterranean Sea in 1201 and may have killed more than a million people. Earthquakes of this magnitude can happen at any time.

In California, the Governor's Office of Emergency Services has set up a warning center to send alerts to state officials during natural calamities. Seismographs constantly monitor California's 10,000 or so tremors per year, most of which are too faint to feel.

Japan has become so earthquake conscious that it has established September 1, the anniversary of the Great Kanto Earthquake of 1923, as Disaster Prevention Day. Residents practice safety techniques such as first aid and mouth-to-mouth resuscitation, fire extinguishing, and breathing from oxygen-filled plastic bags to escape heavy smoke.

At the National Research Institute for Earth Science and Disaster Prevention in Tsukuba, Japan, engineers test and improve the standards for building construction and design. In one instance, atop the Osaka World Trade Center, engineers have placed a computer-controlled sliding weight that, during earthquakes and strong winds, shifts the tower's center of gravity to counterbalance the structure.

Enormous efforts have also gone into the redesigning of columns to support elevated roadways—many of which collapsed during the January 1995 earthquake that devastated Kobe. At Tokyo's Kajima Construction Company, researchers are testing rubber cushions to dampen and flex rigid, brittle structures that otherwise might fracture and crumble in a quake.

To prepare for meteorological hazards, NOAA and the National Weather Service have established warning and forecast centers that monitor such events as severe storms, flash floods, lightning, tornadoes, and hurricanes. Meteorologists at the centers try to track weather events before they strike and claim fatalities. They issue warnings ranging from a few minutes to 48 hours before something happens, thereby giving residents the time they need to take cover or evacuate. These centers help people prepare for some 10,000 severe thunderstorms, 5,000 floods, 1,000 tornadoes, and a couple or so hurricanes that strike the U.S. each year. On average, flash floods claim roughly 140 victims annually, followed by lightning, which kills 93 and injures 300, and tornadoes, which kill about 80 and injure some 1,500. Hurricanes today take a similar toll, but without timely warnings casualties could easily exceed hundreds of victims.

Keeping an eye on hazardous storms, meteorologists at the Storm Prediction Center in Norman, Oklahoma (formerly the National Severe Storms Forecast Center in Kansas City), track supercells—severe, persistent thunderstorms that spawn multiple tornadoes—and issue advisories swiftly. On Palm Sunday, March 27, 1994, for example, after forecasters had steadily launched advisories for 30 hours, culminating in the issuance of a Most Severe

California's Northridge earthquake on January 19, 1994, fractured freeway overpasses and cut off electricity, telephone service, and water and gas supplies. In addition to harming life and property, the quake caused a long-term economic impact. Damage to major commuter routes disrupted traffic for months. To get people and businesses back on their feet, governments speeded up efforts to repair the infrastructure.

Weather Outlook, 18 tornadoes tore through Alabama and Georgia, killing 42 people. Tragically, on that Sunday, one of the twisters ravaged the Goshen United Methodist Church in Piedmont, Alabama, whose falling walls killed 20 worshipers and injured 90 others.

From the 1930s through the 1980s, the number of reported tornadoes soared from 1,685 to more than 8,000 per year—mostly because of improved reporting—yet the number of fatalities caused by tornadoes has plunged from 1,947 to 585. This fatality falloff probably results from improved education, preparation, warnings, and tornado alerts. Those numbers include the deadly outbreak of 1974, in which 148 tornadoes in 13 states killed 315 people, and the Tri-State Outbreak of 1925, when tornado clusters claimed 689 lives in Missouri, Illinois, and Indiana. During the past 20 years, no outbreak has been nearly so deadly, and this fact may be the result of early warnings, better preparation, and a more informed public.

The Storm Prediction Center takes advantage of satellite images, twice-a-day weather-balloon releases, lightning-tracking systems, forecast models, and Weather Service radars to spot trouble when it brews. During the first 6 months of 1994, for instance, 16 killer tornadoes—14 of them solicited official tornado watches—caused 55 deaths.

Tracking wind speed and direction, specialized Doppler radars help forecasters to see inside storms and look for rotation, a telltale tornado sign. The radars detect electrical impulses reflected from rain, snow, hail, and insects. They permit researchers to view cold fronts and airflows over urban centers, and to predict rainfall amounts. Scientists in the VOR-TEX program of the National Severe Storms Laboratory acquire field data on storms from aircraft equipped with Doppler radar and from jeeps beneath frontal clouds.

At airports, Doppler radars have improved warnings of potentially deadly wind shear, downbursts, or microbursts—violently short-lived downdrafts that can cause aircraft to lose lift and crash during takeoff or landing. In 1975, for example, Eastern Airlines Flight 66 and Continental Flight 469 both crashed during thunderstorms. In June, Flight 66 crashed while landing on a runway at New York's Kennedy International Airport, killing 113. In August, Flight 469 plunged into a wheat field during takeoff from Denver's Stapleton International Airport. In July 1994, a microburst may have contributed to the crash near Charlotte, North Carolina, of USAir Flight 1016, which killed 37 passengers. No Doppler radar was present at the airport.

The same necessity for detection and warning applies to hurricanes. In August 1992, meteorologists at the National Hurricane Center in Miami began tracking a storm off the coast of West Africa ten days before it grew into a tropical storm and then became Hurricane Andrew. Using satellite data, airplane reconnaissance observations, and radar, hurricane specialists tracked the storm, providing a warning 21 hours before Andrew devastated Florida's south coast.

Adequate warning enabled more than two million people in Florida and Louisiana to flee their homes. Hurricane Andrew packed winds that gusted to 175 miles an hour and caused a 16.9-foot storm surge. The storm destroyed 80,000 homes and left 55,000 uninhabitable. While 160,000 people were left homeless by the hurricane, 43 Floridians were killed—a tragic number that likely would have been much larger without the storm warning and emergency evacuation measures.

One need only recall the hurricane-bred horrors in the Galveston, Texas, area in 1900, claiming 8,000 lives, or the 1935 Florida Keys hurricane, killing 408 people, to see how many victims hurricanes can claim without warning. By storm prediction or forecasting, the National Weather Service can provide up to 2 or 3 days warning of hurricane landfall, and 12 to 24 hours notice of the possibility of tornado development or conditions leading to the creation of wind shear. Warnings of specific tornado or wind-shear events typically leave only a few minutes for people to take precautions. Meteorologists hope that the new Doppler radar systems will extend warning times for larger tornadoes to 15 or 20 minutes.

Winter weather brings other storm hazards: blizzards and northeasters. In March 1993, when a superstorm hammered eastern North America from Cuba to Canada, the National Weather Service had issued warnings 48 hours in advance of the approaching "storm of historical magnitude." Storm surges, tornadoes, heavy snows, torrential rain, and icy cold blasted coastal cities, washing away 18 homes in Southampton, New York, killing 270 people, and shattering 140 weather records. If warnings had not been issued, the death toll easily could have doubled or tripled. More than one hundred years earlier, in March 1888, a great blizzard took a terrible toll. It blew into the northeastern U.S. without notice, and more than 400 people died.

Because Bangladesh is vulnerable to big storms—seven of this century's nine deadliest cyclones caused more than half a million deaths there—the population now readily accepts and

During the Great Flood of 1993, when the Mississippi River and its tributaries swamped the U.S. Midwest for weeks, local residents and volunteers came together in a remarkable effort to save their homes from destruction. Below, residents of St. Charles, Missouri, a town deluged by floods, struggle valiantly to fill and transport thousands of sandbags for holding back the rapidly rising waters.

ANDREA BOOHER/FEMA

benefits from weather warnings. In 1994, for example, a satellite-linked storm warning system broadcast alerts of a coming cyclone, prompting coastal residents to seek refuge in concrete storm shelters.

While the 1994 cyclone tragically claimed several hundred lives, that number amounted to a fraction of the 139,000 fatalities caused by a cyclone in 1991, when a 20-foot-high storm surge swept over densely populated coastal and island villages. In a 1970 cyclone 300,000 people perished similarly. Today's storm warning system and concrete storm shelters, built on low-lying islands with the help of foreign governments and private relief agencies, may help soften the blow when cyclones strike again.

SCIENTIFIC INQUIRY

Scientific inquiry into the causes of nature's deadly events helps us not only to prepare for future ones but also to understand how past disasters unfolded.

Consider the demise of the Akkadians, whose empire 4,300 years ago vanished in less than two centuries. Archaeologists sampled 80 centuries of soil from the northern part of the former empire and found that high sand and dust levels during the years of decline were likely caused by droughts and dust storms. Because similar disasters befell Greece, Egypt, Palestine, and the Indus Valley at about the same time, researchers ask whether a sudden climate shift struck these cultures in the 3rd millennium B.C.

At the National Ice Core Laboratory in Lakewood, Colorado, scientists study glacier-ice cores taken from the Greenland ice sheet. Holding dust, volcanic ash, pollen, and air bubbles, the ice preserves a 250,000-year climate record. Analyses of it, and ice from Antarctica and the Andes, indicate that the climate shifts more frequently and more radically than scientists had once thought.

Researchers use computer models to forecast future climates, putting in equations for the earth's size, rotation rate, seasons, and topography, as well as for water, air, and energy transfer. Then they vary factors such as cloud cover, volcanic eruptions, and ocean currents and see what the models do. Scientists also use them to explain and track changes in Pacific surface temperatures. Of critical importance is understanding El Niño, an interaction between the atmosphere and the ocean that can disrupt weather around the world. Predicting its onset has great practical value.

PETER AARON/ESTO

Researchers at the National Severe Storms Laboratory in Norman, Oklahoma, and other sites are working to understand storm mechanisms, especially why and how tornadoes form. If they can understand the causes of tornadoes, for example, then they can recognize their development at the earliest stages and quickly issue warnings.

With this goal in mind, tornado-chasing scientists in vans fitted with Doppler radar probe tornado-spawning supercells while radar-mounted aircraft swoop in to image the storms from above. In June 1995, scientists obtained detailed radar images within two miles of a tornado. Forecasters believe they will soon be able to spot "signature" disturbances to identify storms with the potential to create tornadoes.

To probe the earth's interior and movements of the crust, researchers use computer-aided laser-ranging systems, satellite data, and other sophisticated geophysical devices. A constant background of earthquake "noise" permeates the planet's crust, and using signature

New York's 59-story Citicorp Center (left) counters wind-triggered oscillations with a "tuned mass damper"—a computer-controlled, sliding concrete block in the tower's crown. If the tower sways more than a foot per second, hydraulic arms (opposite) push the block in the opposite direction, damping oscillations 40 percent.

of hazardous locations. In 1994, volcanologists detected ground movements and earthquakes below volcanic peaks near Rabaul, Papua New Guinea. Now they monitor movement of sub-surface magma that may signal an eruption as massive as one occurring there 1,400 years ago.

To find ways to help buildings withstand earthquakes, the National Institute of Standards and Technology in Gaithersburg, Maryland, and the Earthquake Engineering Research Center at the University of California, Berkeley, are testing new materials and structural designs in machines that simulate earthquakes.

Engineers are using "passive solutions" in building design to help mitigate the effects of earthquakes. Buildings are set on alternating layers of rubber and steel plates or sliding bearings to reduce shaking and dissipate energy. San Francisco's Transamerica Building, for example, takes advantage of a pyramid design and a heavy mat foundation.

Engineers in the U.S. and Japan also test "active" systems that use computers to sense vibrations and shift counterweights to counter-act movement. New York's Citicorp Center uses a 400-ton concrete block floating on oil film to reduce the swaying of the building in the wind to a mere 6 inches. And at the top of the Osaka World Trade Center, a massive counterweight slides to shift the skyscraper's center of gravity during quakes or wind shear.

vibrations in the sounds, scientists can build models of internal structure. Besides listening passively, geologists generate shock waves with directed explosions, then track reflected vibrations, carefully decoding their signatures to map subsurface structure and rock formations. The behavior of the artificially generated seismic waves helps them model structural features that would otherwise remain undetectable.

Sensitive monitoring also benefits residents

SEVERE STORMS
RAIN, SNOW, WIND

by Jeff Rosenfeld

INTRODUCTION

In January 1996, a blizzard paralyzed the East Coast of the United States and with it much of a busy nation. A foot of snow fell in Georgia and Maine; in between, nearly two feet buried the wide boulevards of Washington, D.C. The federal government closed its doors for almost a week, and a backlog of 200,000 applications mounted at the passport office. Blood supplies ran dangerously low at the Red Cross. Farther north, in Philadelphia, a city-record 30.7 inches fell. The *Inquirer* missed daily publication for the first time in its 166 years. Trash collection stopped for nearly a week. Letter carriers missed their rounds.

Along the Northeast corridor, abandoned cars littered the highways, and frozen switches crippled the railways. Snowbound airports made the storm a national disaster. Airlines canceled 10,000 flights nationwide, stranding thousands of travelers as far away as hot and sunny Los Angeles. Up to 2 feet of snow fell at New York's airports, where drifts were 20 feet tall. When runways reopened after two days of plowing, one jet bound for Tokyo taxied into a snowbank. The 264 angry passengers spent more than 7 hours stuck on board without ever taking off.

In New York City, nearly a million children took a day off as snow shut down public schools for the first time in 18 years. Scarf-swaddled pedestrians in midtown Manhattan posed for pictures in the middle of Broadway, a great white way. On Fifth Avenue, tourists stranded with nowhere to shop climbed a 12-foot pile of snow to survey the eerie quiet. "It's bad and it's beautiful," one observed.

The nerve center of the United States had indeed gone numb in the snow, as more than a hundred people died and billions of dollars in losses quickly accumulated. The silent aftermath was a reminder that it is not the city but the weather that never sleeps. The atmosphere's business must go on. That business—to redistribute heat and energy—meets a never-ending planetary need: As the sun heats the surface of the earth unevenly, differences in temperatures arise. Through rain, sleet, hail, and snow, the weather settles those differences.

The tireless troposphere, the bottom layer of the atmosphere, is saddled with most of the work. Hot air in the tropics rises and moves slowly poleward, finally descending between 20° and 35° latitude, where it contributes to desert conditions. Meanwhile the denser, colder air at the Poles sags toward the tropics. In the otherwise temperate latitudes, warmth meets cold and the differences are brought to a head. Extreme storms are born—blizzards, thunderstorms, and tornadoes—in the turbulent completion of the atmosphere's task.

These storms not only sort heat but also deliver water. Workhorses in a perpetual cycle, storms lift moisture from the oceans, lakes, and rivers and eventually deposit it at higher elevations, where it runs downstream, changing the landscape and slaking our thirst.

A storm like the January blizzard mobilizes millions of cubic miles of air and water vapor to complete these urgent global tasks. Every endeavor underneath, no matter how great, yields to the weather's demands. In this century, however, we have emerged from beneath storms to anticipate, study, and admire them at work. With computers, meteorologists simulate

New Yorkers dig out from under nearly two feet of snow deposited by the Blizzard of '96. And even after the record snows ended, residents faced more weather woes: Subsequent rains and thaw proved just as treacherous, forcing 125,000 people to flee their homes when floods ravaged the Mid-Atlantic states.

PRECEDING PAGES: Darkened by the Kansas soil it ingests, a tornado crosses Wilson Lake. Winds in the most intense tornadoes reach nearly 300 miles an hour in a hollow spiral that can extend a mile across.

weather in the safety of virtual worlds. With radar and lasers they probe tempests from afar. With airplanes, they meet weather on its own turf, and with satellites they give us a dazzling view high above the storms that besiege us.

Catching a flight out of Washington days after the Blizzard of 1996, one lawmaker, relieved by the lofty perspective, felt the frustrations below melt away. "Once you're above the clouds," he said, "it's great."

STORM FRONTS

During World War I, Norwegians suffered from famine. Unable to import grain, the people were desperate to improve food production. But farmers need weather forecasts, and wartime secrecy blacked out information about storms moving from the North Sea and the Atlantic. Vilhelm Bjerknes, a pioneering atmospheric scientist, tackled this problem by establishing a network of observers linked to his Bergen headquarters by telephone.

The result was a revolution in meteorology. With the data, Bjerknes's son Jacob formulated a theory that made sense of the extratropical cyclone—the stormy circulation around low-pressure areas in the midlatitudes. His depiction of weather as a struggle between warm and cold air along fronts was reminiscent of the battle fronts of the First World War.

Jacob Bjerknes knew that cold, dense air spreads equatorward from the Poles and meets warmer, lighter air from the tropics. The air does not mix easily and instead faces off in a convergence zone of great temperature contrast called the polar front. The front is an atmospheric battle zone encircling the earth. Here air masses, mounds of uniform air, vie with one another. Covering hundreds of thousands of square miles, they take on the character of their origins: Air from the Gulf of Mexico, for instance, is moist and warm; air that pools over northern Canada is dry and cold.

Where air masses collide, lows evolve and deform the polar front. Low-pressure centers draw in air and, because of the earth's rotation, establish cyclonic circulation. In the Northern Hemisphere, flow is counterclockwise; warm air invades northward east of the low, while cold air invades southward west of the low. "The warm air is victorious east of the center," the elder Bjerknes explained. "The cold air…makes a sharp turn toward the south, and attacks the warm air in the flank." The vanguard of cold air is a cold front; advancing warm air is a warm front.

These fronts extend from a low and pivot around it like the hands of a clock (winding backwards in the Northern Hemisphere). Warm air between the fronts is often called the warm sector. This sector disappears when the faster cold front catches up to the warm front, forming an occluded front. An occluded storm has reached its peak; it soon dies as warm and cold air finally mix, eliminating the temperature contrasts that supply energy.

With winds converging toward the low, the frontal model explains why winds shift as fronts pass. Fronts also account for the fast temperature changes a storm causes. They explain precipitation patterns, because they lift air. In the atmosphere, if a bubble of air rises, there is less air above it, so the weight of air above it decreases and the pressure of the bubble drops. The bubble expands and cools as a result. (The reverse, a warming by compression, occurs when air sinks.) If rising air cools enough, the water vapor in it condenses, forming clouds.

In a classic cyclone, the warm front at the surface advances, and light air climbs over the colder air north and west of it. The ascent is sometimes less than 30 feet a mile, and depends on temperature contrasts and winds. The resulting clouds spread precipitation along and ahead of the warm front. After the front passes, temperatures rise. As the cold front advances, dense air behind it wedges under the warm sector. The

Clouds heavy with rain hang low over the Great Plains. Where atmospheric fronts sweep across terrain, air masses collide, sculpturing local storm clouds. This basic mechanism of weather—the clash of hot and cold air along boundaries—helps meteorologists explain the atmosphere's daily traveling show of shapes, colors—and danger.

wedge is twice as steep as the warm front, steeper if the cold front advances quickly. The sharp lifting initiates cumulus clouds or thunderstorms in a narrow band of precipitation. After a cold front passes, the sky clears and cool winds blow. In 1995, a front in Idaho dropped the temperature from 80°F to 35°F in 4 hours.

The frontal model explains different types of precipitation. Ice crystals grow in rising, cooling air, eventually growing too heavy to stay aloft. If ice crystals fall through warm air, they may melt into rain. If they descend through cold air, they fall as snow. In the three-layer sandwich of an occlusion, ice crystals might fall through warm air and melt, then fall through cold air near the ground and refreeze as sleet. Or, if the surface cold layer is thin, the melted ice may refreeze on impact, causing freezing rain.

A quick check of daily weather maps shows that the classic frontal model is an ideal usually muddied by circumstances. The structure of frontal storms is complicated by the effects of mountains and oceans, and scientists still argue over the true profile of occlusions. Some storms exploit midlatitude temperature contrasts and then seemingly abandon the model altogether. A northeaster can develop an eye like a hurricane. An Arctic storm may start from a frontal boundary and then run off with a hurricane-like eye. The model does not explain these findings, but it remains a reasonably reliable guide to forecasting storms like big blizzards.

STORM MOVEMENT

In 1743, Benjamin Franklin missed a lunar eclipse when a storm struck Philadelphia. When he learned that his brother in Boston had seen the eclipse, but had also seen a storm later, Franklin did a little research by mail and became the first person to prove the movement of storms. Following his lead, today's forecasters watch lows and their fronts move thousands of miles. But figuring out where rain or snow will fall at what time is still a tricky business, involving modeling the atmosphere on the world's fastest supercomputers.

Storms follow a train of waves in upper-level winds—the key to today's forecasts—discovered in the 1930s by Swedish meteorologist Carl-Gustaf Rossby. Winds are caused by pressure differences, and because of the passage of lows, winds shift a great deal near the surface. But high-altitude winds generally move eastward between 30° and 60° latitude. These winds result from the contrast between warm air near the Equator and cold air near the Poles, which produces pressure differences that increase with altitude. The region of greatest cross-latitude temperature contrasts migrates poleward in summer and equatorward in winter, and hence so do the upper-level westerlies.

Free of surface friction, winds aloft regularly reach 150 miles an hour. In World War II, pilots discovered such winds at 25,000 to 40,000 feet. Bombers slowly made westward headway toward Japan or England, but flew eastward at blazing speeds—if they didn't run out of fuel first. The high wind turned out to be a core of peak flow within the westerlies. Now called the jet stream, this high-speed core largely determines where and when storms will strike.

The usual positions of the Rossby waves in the jet stream over North America trace a number of the typical tracks storms follow. Alberta clippers are compact lows that often appear in the lee of the Canadian Rockies and head southeast, towing lots of polar air behind them and bringing a dusting of snow. Panhandle lows from Texas gallop northeastward toward the Ohio Valley and the eastern Great Lakes. Gulf lows are disturbances that stray into warm waters south of Texas, often breaking through Florida and sweeping up the East Coast. If they remain over water, becoming coastal lows, they are often called northeasters for their strong onshore winds north of the pressure center.

Coastal storms are a special breed. Like all extratropical storms they feed on temperature contrasts, and the contrast between cool and warm ocean currents can intensify them. For instance, the convergence of the cold Labrador Current and the warm Gulf Stream frequently ignites coastal lows off the Carolina coast.

In many cases, North America's lows originate over the Pacific Ocean (just as European cyclones may start as disturbances in the U.S.). But the Pacific lows compress vertically and spread out over the mountains of the West, weakening and sometimes disappearing from weather maps until they reach the High Plains. Through all of the mutations caused by traversing water and mountains, however, storm circulations can be quite hardy. Satellite film showed one wave-like disturbance in upper-level winds circling the earth twice on a 57-day journey, sowing storminess in various incarnations.

Between 20,000 and 40,000 feet, long waves— about 4 to 6 spanning the earth—settle into the

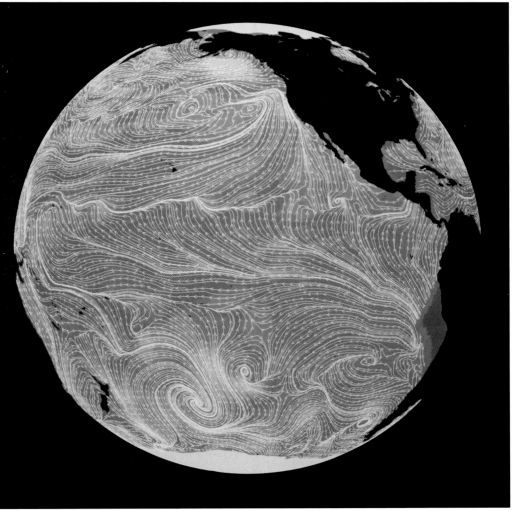

Centuries ago, Benjamin Franklin read his mail to discover the movement of storms. Following his lead, meteorologists since 1961 have used images from satellites orbiting the earth to track weather disturbances. Computer processing of data from NASA satellites produced this snapshot of global airstreams: Blue and gray denote light winds; red and yellow designate strong winds. Arrows curl into the cores of storms traversing the South Pacific Ocean and the waters just south of Alaska.

westerlies. They move slowly, but forecasters are still often bedeviled by shifts in these waves. Thus, forecasts of individual storms more than a few days ahead can be very uncertain. Not surprisingly, a meteorologist with a simple computer model of the atmosphere discovered the essence of chaos, a sensitive dependence of an evolving situation on its initial conditions. That meteorologist, Edward Lorenz, helped spur a host of chaos studies.

Forecasters have recently begun using a new technique, called ensemble forecasting, that at least reveals the effects of chaos on the atmosphere. Meteorologists gauge the uncertainty of their forecasts by varying slightly the initial conditions fed into the computer models. If the results vary widely, then the forecasters know that chaos is rendering their projections less reliable. If the forecasts vary little, then the forecast is more reliable.

Forecasters also consider jet-stream developments when they assess the intensification of a storm. Jacob Bjerknes expanded the frontal model into a synergy of winds at the surface and at high altitudes. Any upper-air flow that helps lift surface air can help intensify a storm. For instance, diverging or accelerating airstreams create a partial void that brings air up from the surface, intensifying a low. Also, differing side-by-side airstreams or counterclockwise waves can encourage surface air to spin around a low. In all, air tends to sink upstream from a southward bend, or a low-pressure trough, and rise downstream from it. These upper-air developments figure prominently in forecasts of an extratropical cyclone's development. On March 9, 1993, four days before snow first fell, meteorologists had computer outputs in hand that predicted a volatile interaction of storm tracks and upper-air divergences. A superstorm would hit the East Coast.

BLIZZARDS / SUPERSTORM '93

The Superstorm of 1993 disrupted the lives of a hundred million people. On the weekend of March 13-14 some motorists slept in their snowbound cars beside desolate, windy roads. Three thousand others overnighted on benches and floors in Atlanta's Hartsfield Airport. Millions stayed home, marooned in the dark cold of power outages.

One man in upstate New York died when a slide of snow buried him as he tried to free his car from a drift. Another died in an avalanche in his own backyard. Four people died in Florida when their boats sank in 80-mph winds and thrashing waves. Others froze to death outdoors. In all, the storm claimed 270 lives. Untold numbers of people later succumbed to heart attacks trying to shovel the snow.

They were victims of an atmospheric circulation covering three million square miles, a confrontation of warm and cold, wet and dry air that exceeded most of the superlatives in its mammoth path. The storm sank a 200-foot freighter in the Gulf of Mexico, then sank a 600-foot freighter off Nova Scotia, drowning 33 men in 65-foot waves. The waves also pushed aside dunes and ate coastal houses in Florida and Cuba. The winds included a tornado that left 5,000 people homeless in Reynosa, Mexico. Far-flung extreme gusts hit 109 miles an hour in Dry Tortugas and 131 miles an hour in Canada.

Snow fell over an area twice as large as the fallout from the Blizzard of 1888, with 5 inches in Florida and 16 in Ontario. The storm buried the East with as much water as the Mississippi River carries in 40 days and 40 nights.

The 1993 storm started as a pack of thunderstorms in the western Gulf of Mexico, but was joined by a wet storm from the Pacific and by a minor snowstorm from the Arctic. The three disturbances met on Saturday, the 13th, in the Southeast. Unlike many coastal storms, which intensify by pushing against an inland anticyclone, the superstorm acted more or less alone. High winds were generated by the contrast between a hurricane-like low pressure in the middle and the storm's surroundings.

The superstorm benefited from the typical winter position of the polar front and upper-level winds. After weeks of long nights and extensive snow, the cooling of the Northern Hemisphere pushed the polar front far south, often over the Gulf Coast. The winter chill also amplified the temperature contrast across the front. The result was a volatile storm production zone on the border between land and sea.

A trough in the upper-level westerlies brought high-speed air looping south in the central states and then north in the East. Upstream from the trough, the jet stream charged southeastward at 165 miles an hour. Cold air from western Canada spread southward, steered by the high-level winds. A fast current of air over the Gulf of Mexico, a southern branch known as the subtropical jet, helped feed air from the Gulf into the southern states. This pattern had prevailed much of February, bringing frequent storms to the East, but had disappeared briefly in the week before. Weather forecasters plugged upper-air profiles and surface conditions into their mathematical models and let the computers calculate future conditions step by step. Already on Tuesday, March 9, the result was an incredibly big weekend storm over the East.

Satellite imagery crystallizes the swirling beauty of the March 1993 superstorm. The thick cold front, stretching from Canada to Cuba, set record lows from Maine to Alabama. Behind this boundary, cold air spills over the Atlantic Ocean and creates arcs of low clouds, while under clear skies the East glistens white with a blanket of new snow.

At sunset Thursday, the expected converging surface air emerged from Mexico; large areas of the Gulf were ablaze with lightning. Early on Friday weather satellites spied tall clouds growing southeast of the low, in the usual warm sector of the low-pressure system. The storm strengthened and by evening had a low enough central pressure to sustain 77-mph winds off Louisiana. Rain spread north of the low as warm Gulf air rode over the cold air spilling off the coast. The storm was drawn into circulation over the United States not only at the surface, but also above. Supercomputers at the National Weather Service ran several prediction models; they were nearly unanimous in pointing the storm toward the Carolinas the next day. The jet stream would grab the low and fling it northward along the East Coast.

Upper-level winds gave the storm more than guidance: They gave it healthy temperature contrasts, pulling in cold air from the north and pushing warm air into the southeastern states. Aloft, the curving flow prescribed a cyclonic twist over the Gulf as the storm developed, and accelerating and diverging airstreams caused lift at the surface.

The unified storm exploded onto the landscape. Its shield of precipitation—largely snow—unfurled across the Southeast. On Friday morning the central surface pressure was an undistinguished 29.54 inches, but by evening it was 28.35 inches and within the range of hurricane strength.

Off Florida the storm acted like an oversize hurricane. Gulf waters rose as much as 12 feet as the low passed near Tallahassee in the dark hours of Saturday morning. Forecasters had not anticipated a great surge, which was boosted by topography and the storm's size. In southerly winds ahead of the low, waters piled into Apalachee Bay, south of Tallahassee. Then as the low and its cold front passed, winds turned into 50-mph southwesterlies, releasing the waters that overwhelmed Dekle and Keaton Beaches. The surge, which was more than a hundred miles wide, was far bigger than that of a typical hurricane. The waves dashed houses into the water, killing people trapped inside.

The superstorm also sprouted a squall line in the warm air ahead of the cold front. This extratropical-cyclone feature extended over the Gulf of Mexico. The thunderstorms sped at 70 miles an hour across Florida and Cuba. At midnight they hit the Tampa area; in about two hours, eleven tornadoes and fierce downdrafts blasted the state. In Dade County a woman died as her trailer was flipped by 80-mph winds; in Cuba, 1,500 houses were destroyed.

The squall line spent itself over the cool Atlantic, but the low turned northward with the upper-level flow. On Saturday, the winds in North Carolina shifted to the southwest as the storm moved northward. Water in the shallow sounds along the coast shifted eastward and was pinned against the barrier islands by winds up to 83 miles an hour. Pamlico Sound rose nine feet along Cape Hatteras. Hundreds of homes were ruined at Long Beach. Four feet of water stopped firefighters in Rodanthe: They looked on helplessly as the post office burned.

Farther inland, the problem was snow and cold. By midday, snow fell simultaneously in New England and Mississippi with the low in between, in North Carolina. A foot of snow fell in Atlanta, Georgia. Drifts were 12 feet high in Steubenville, Ohio, and 21 feet high in Boone, North Carolina. In the southern Appalachians, at least 27 people froze in wind chills of minus 20°F. Helicopters rescued hundreds of campers from the forests. More than a dozen motorists got stuck in snow near High Knob, Virginia, and waited 28 hours for rescuers to reach them. Elsewhere people died huddled in their cars; others died searching for shelter.

Whereas many blizzards in the Rockies accumulate tremendous snows in powdery falls, the superstorm was noteworthy for its wet snow. An average foot of snowfall might melt into little more than an inch of water; in the cold, dry Rockies, a powdery pile two or three feet deep might only amount to an inch of water. In Asheville, North Carolina, every foot of snow equaled nearly three inches of water. A typical suburban driveway lay under four tons of snow. With widespread heavy snow and sleet toppling power lines, three million people lost power and hundreds of roofs collapsed.

Precipitation in a coastal storm is very sensitive to the low's position. This storm stayed far enough inland that rain fell in the coastal plain: Not enough cold air carpeted the area. The passing frontal structure created a mix for some areas: South Carolina got rain, sleet, and snow. So did Washington, D.C., where snow changed to sleet and rain during the day on Saturday and turned to snow again at night. In central New Jersey, two inches of sleet topped a foot of snow. The mix congealed into a thick ice that proved stubbornly resilient to salt.

The largest 24-hour snowfalls in a decade piled up in Providence, Boston, and Hartford,

Icy winds winding about a deep low-pressure core as powerful as a warm-weather hurricane made the Superstorm of 1993 more than just another snowstorm. Foaming seas surged inland, battering shoreline communities from Mexico to New England, including Long Island's Southampton Beach (below), where high water swept away 18 homes.

and nearly two feet of snow buried Long Island. Over White Plains, New York, the storm reached its lowest pressure, 28.38 inches. Extratropical lows can be even stronger. A 1992 storm that hit England had a pressure of 27.64 inches over the North Atlantic, for instance. Still, such storms could not approach the superstorm's size, nor could the supertyphoons of the Pacific. A hurricane's maximum winds might be 30 miles from the center: The superstorm's winds spread over a 200-mile radius.

On Sunday, the skies cleared and winds settled down to a whisper across the snowbound South. Behind the front, cold air set record lows ranging from minus 12°F in Caribou, Maine, to 31°F in Daytona Beach, Florida. The low in Birmingham, Alabama, bottomed at 2°F. Three-quarters of Georgia's peaches were frozen. In all, the storm had cost six billion dollars.

Nine months later, a wave of births along the East Coast added thousands to the number of lives touched by the Superstorm of 1993.

BLIZZARDS / LAKE-EFFECT SNOW

Thanksgiving Day in 1903 began as a pleasant holiday, with sunshine over much of the Chicago area. But around noon the phone rang, and Weather Bureau meteorologists downtown got a shock. "When will it stop snowing?" asked a university official on the south side of town. "What snow?" asked the stunned forecasters.

That day northerly winds picked up moisture over Lake Michigan and deposited 14 inches of snow on South Chicago while just a few miles inland and north, not a trace of precipitation fell. It was one of the most bizarre of a long line of lake-effect snows in the region. More recently, in December 1995, cold air from Alaska traversing the lakes delivered a record 61.7 inches in Sault Ste. Marie, Michigan, and a one-day record of 37.9 inches in Buffalo, New York.

Lake-effect snowstorms make the Great Lakes the snowbelt of the United States. While high elevations in the West regularly get more snow and East Coast snowstorms sometimes cover larger regions, the lakes develop frequent localized snowstorms with great vehemence.

Averaging more than 110 inches of snowfall a year, Syracuse, New York, near Lake Ontario, is the snowiest major city in the United States. But the lake effect is far more intense elsewhere. In the winter of 1976-77, for instance, while Syracuse got its usual snows, totals were astronomical southeast of Lake Ontario. Old Forge had 408 inches, yet lost its Great Lakes regional snowfall record to nearby Hooker, where 467 inches were measured.

On the northeastern shore of Lake Erie, Buffalo had 199 inches, with 53 consecutive days of snow. To top it off, a blizzard on January 28 blew in on a cold front at midday, stranding workers at their offices in downtown Buffalo as 85-mph gusts built 25-foot drifts with 12 inches of new snow. Twenty-nine people died—some in their cars, buried in snow. Buffalo was the first place to be declared a federal disaster area because of snow.

Had Lake Erie not frozen over completely in midwinter, Buffalo surely would have gotten even more snow. That is how the lake effect works. Cold dry air traveling many miles over Canada sweeps across the lake, often after a low has arrived in eastern Canada. The air is heated from below, and snow-bearing clouds form as water evaporates from the lake surface, especially when the air-lake temperature difference is about 20°F or more. The heavy clouds dump their moisture in a hurry, as snowfall. When the lakes freeze, evaporation is shut down for the season and the lake-effect storms hibernate.

Lake-effect snows in the region usually begin in mid-autumn and are often initiated by shore breezes—shallow, cold airstreams from land that converge and rise in small areas of low-pressure in the middle of the lake. The resultant clouds form a snow band that remains until a larger weather system dislodges it. Some of the snow bands line up parallel to cold northwest winds; others may be parallel to shore. The sun may shine on either side of a five-mile-wide band, but underneath, as the clouds blow inland, several feet of snow can fall.

Great Lakes storms are known for more than just the snow they bring. On November 9, 1913, a Canadian blizzard squeezed in from the west while a hurricane-deep low-pressure center intensified over Ontario. The long fetch of

Snow stopped motorists in their tracks in Buffalo, New York, when a five-day blizzard raged during January 1977. At last the sun came out, and drivers returned to buried roadways to dig out their abandoned cars (above). Lake shores often form the hub of a hard winter until a sheet of ice—like that forming in Chicago during a 1994 cold wave (below)— seals off the water supply necessary for lake-effect snowstorms.

northerly winds, topping 70 miles an hour, drove huge waves into the south end of Lake Huron. Nineteen ships went down and 20 more were dashed onto rocks as 250 sailors drowned in churning, frigid waters. One freighter was later found floating upside down in the middle of the lake. Bodies washed ashore weeks later.

The storm and its long northerly fetch over water also produced a memorable snow in Cleveland, Ohio: 22 inches with 8-foot drifts as winds hit 50 miles an hour. But the ice dragged down miles of telegraph and power wires, cutting off the city. Many days passed before Clevelanders, or anyone else, realized the extent of the tragedy on the lake.

"The wind was straight in my face and beat so in my eyes that I couldn't see a rod before me.... I stumbled along, falling down at almost every step, burying myself in the snow when I fell, struggling frantically up only to sink deep down again.... Every time I fell down, I shouted and cursed and beat the snow with my fists...." This is the account of a New Jersey man trying to walk a few blocks to a friend's house in Jersey City Heights on "Blizzard Monday," March 12, 1888.

The blizzard blanketed a broad area from northern Virginia to Maine. It isolated Washington, Baltimore, Philadelphia, New York, Boston, and other cities, and buried hundreds of towns and villages. Over southeast New York State and southwest New England, snowfall ranged from 30 to 50 inches, with wind-driven drifts of 30 to 40 feet. For two days, gale-force winds and record low temperatures assaulted the most populous part of the nation.

Men, women, and children died on city streets, in country fields, in stalled trains, and on ice-choked vessels. More than 400 people died, 200 in New York City alone. The blizzard buried trains all over the Northeast, in some cases marooning passengers for a week or more. Many of them left the cold, drafty cars to find food and shelter; some perished. The storm also sank, grounded, or wrecked 200 vessels from Chesapeake Bay through New England waters. At least a hundred seamen died in the "Great White Hurricane." Although more than twice as much snow fell farther north and east, the blizzard became a legend in New York City.

Saturday, March 10, was a warm, sunny day in New York. The weather "probabilities" in the *New York Times* promised "warmer, fair weather" for the next 24 hours. No one took much notice when rain began falling on Sunday afternoon. The rain turned to snow at 12:10 a.m. on Monday and continued through the day, falling at the rate of an inch an hour until 3 p.m. It was driven by fierce northwest winds that peaked at

CORBIS-BETTMANN

38 miles an hour, according to new research. Temperatures plummeted to 8°F by midnight, and below 5°F by early Tuesday morning.

Most working people first felt the storm's fury when they tried to go to work Monday morning. Many were blown off their feet and had to crawl on hands and knees. Others were sprayed with flying glass from shattered windows or felled by flying signs or falling chimneys. The snow forced drivers and passengers to abandon stalled horse-drawn streetcars and steam-driven elevated trains. Commuters jammed platforms waiting for trains that never came.

MUSEUM OF THE CITY OF NEW YORK

As depicted in this engraving (opposite), wind and snow felled New Yorkers in the Blizzard of '88. Afterward, Wall Street bankers and brokers inspected damage to telegraph and telephone lines (left). The storm downed lines throughout the Northeast, cutting it off from the world. As a result, many cities soon put wires underground.

More than 20 letter carriers were found unconscious in snowdrifts in Brooklyn, where many poor families had to be carried from icy, unheated hovels to police stations that were overflowing by noon. Meanwhile, bars, hotels, private homes, and public buildings soon were stuffed with people seeking shelter from the storm.

With horse-cars, trains, steamers, and ferries immobilized by the storm, some of the workers tried to return home by walking across the Brooklyn Bridge. They had to pull themselves across hand over hand, clinging blindly to the railing while fierce, turbulent windblasts buffeted them and blew their hats into the East River. Police finally closed the bridge, fearing someone would be blown off, or die of exposure or exhaustion.

By nightfall, New York was an eerie, arctic wasteland. The streets were deserted. Men stranded in the city knew nothing of families at home, where wives and children could only guess the fate of father or husband. Some of the stranded spent the night in jail, while a number of wealthy businessmen marooned in the financial district shared quarters with skid-row inhabitants. The snow tapered off Tuesday morning, though several more inches fell that afternoon and on Wednesday. Meanwhile, sleighs, skis, and snowshoes were common sights on city streets. One woman remarked it was strange to see sleighs sliding over snowbanks at the level of second-story windows.

The blizzard's rare combination of prolonged gale-force winds, a severe cold wave, and an almost unprecedented snowfall indelibly imprinted the storm in the memories of all the people who lived through it. Indeed, 41 years later, two still-awed survivors organized the "Blizzard Men and Women of 1888," who for more than 40 years met on March 12 to commemorate and tell tall tales about the great Blizzard of '88.

Patrick Hughes

THUNDERSTORMS

High above a grassy meadow, a hawk circles slowly for hours. Its wings seem still, its flight effortless, as it waits patiently for its prey.

The path of a hovering bird traces the invisible workings of the atmosphere. In light winds, delicate cells of convection bubble up from hot spots on the earth's surface. Birds seek these updrafts and circle within them to conserve energy. When the wind blows harder, the rising air spirals along the wind's direction, and the birds drift straight ahead atop the cells.

Sometimes the cells of air are bigger—too violent for bird or beast. In 1959, Lt. Col. William Rankin, USMC, had to bail out of his fighter jet into a thunderstorm towering 45,000 feet above Norfolk, Virginia. Rankin quickly numbed in the minus 70°F cold as he plunged into the roiling storm at more than 100 miles an hour, his ears bursting and eyes bulging in the thin air. Deep in the "soft, milk-white" cloud, his fall was barely perceptible; up was indistinguishable from down. Then baseball-size hail hammered his body, his helmet saving him from concussion. When Rankin's parachute opened at 10,000 feet, the thunderstorm shot him upward with jarring force. For 40 minutes he rocketed up and down in the currents of the storm's convection cells, "an angry ocean of boiling clouds...spilling over one another, into one another, digesting one another."

Thunder vibrated through his body in a deafening roar, and huge sheets of bluish lightning laced the air around him. Rankin was swimming in air thick with tons of water: "I thought I would drown in mid-air...several times I had held my breath, fearing to inhale quarts of water." The marine landed in a North Carolina forest, lucky to survive.

A series of simple convection cells is all that a thunderstorm needs to flush a narrow Colorado canyon, bury a Texas county in hail, or flatten a million trees in the Adirondacks. The power for storms, like all convection, comes from the sun.

Sunshine has little direct impact on air temperature. It is mostly the surface of the earth, warmed by the sun's rays, that heats the air from below. Within the atmosphere's bottom layer, or troposphere, the air temperature normally decreases as the altitude increases, facilitating up-and-down motions and leading to the formation of a variety of interesting weather. By contrast, the temperature within the stratosphere is nearly constant, and the air is so dry that clouds do not form. In the troposphere, where the action is, slowly sinking air occupies huge areas with good weather, whereas rising air tends to be concentrated near storms, large and small.

Rising bubbles of air cool by expansion; if they cool enough, water vapor begins to condense into water droplets or ice crystals, and the air bubbles then become visible as puffy, white clouds known as cumulus. Often cumulus clouds are signs of weak convection, of air that perhaps begins less than one degree warmer than its surroundings and rises slowly, much like an old elevator. Air bubbles may stop, mix with cooler air, and abandon their pursuit of altitude.

But this is not the fate of a cell that uses its moisture advantage. Water and ice formation in the cloud releases a tremendous amount of heat

In less than an hour, a severe thunderstorm far outstrips its humble origins as a pocket of rising air. Here, a Texas storm rings its core of precipitation with a cloudy base that signifies air rising and condensing into the updraft tower above. A single, volatile rumbler like this thunderstorm can threaten an area as much as 25 miles across with lightning, floods, hail, and high winds.

and counteracts the expansional cooling. Air cools about 5.5°F for every thousand feet of elevation gain; it cools less than 4°F for every 1,000 feet as water condenses. If enough water condenses and releases heat, a bubble may get so much warmer than its surroundings that it climbs wildly. This is an unstable atmospheric scenario, and if it extends all the way to the top of the troposphere, the atmosphere is ripe for giant cumulonimbus clouds that dwarf city skylines. At any given moment on earth, some 2,000 tight collections of cells are raging as thunderstorms—the tallest of them twice as high as Mount Everest.

A typical thunderstorm may last an hour, rising first as cumulus clouds coalesce into a mile-wide collection of cells called a *cumulus congestus* and building upward at about 1,000 to 2,000 feet per minute. When the air inside tops 20,000 feet in a severe thunderstorm, it can be warmer than its surroundings by 7°F or more. In the meantime, as the cells build to well above the freezing altitude, ice crystals start to form in the cloud and grow to about a million times the size of a single cloud droplet. Then, beginning a maturity that might last a half hour, the storm does the two things it does best: It drops heavy precipitation and generates lightning. The icy proto-raindrops fall when they are too big for the updrafts to hold, creating downdrafts in a cloud tower perhaps ten miles wide at its base.

The thunderstorm changes its appearance at maturity. A precipitation shaft extends from its base, and the top of the storm, formerly looking hard-edged like cauliflower, begins to blur with ice. At the top of the troposphere, updrafts may have enough momentum to climb as much as 9,000 feet into the stable stratosphere before falling back into the troposphere where they belong. Ultimately most of the ice at the top of the storm spreads in upper-level winds, giving the cloud a characteristic anvil-shaped top that can stretch beyond the horizon.

Eventually the cells exhaust supplies of warm surface air and dissipate—sometimes the anvil's shade helps cool the storm's own heels. Updrafts weaken, precipitation abates, downdrafts weaken, and the cumulonimbus may fizzle out in half an hour. A typical thunderstorm cloud contains convection bubbles in all stages of development—cumulus, mature, and dissipating. Small cumulus often may grow on the flank of a mature thunderstorm, becoming a new storm as the older one dissipates.

Given the importance of surface heating, thunderstorms are understandably common in tropical regions. Elsewhere they are usually a summertime phenomenon. Feeding off daytime insolation, thunderstorms also tend to peak in the afternoon or evening. The open ocean and the eastern Great Plains are exceptions. At night, the ocean surface remains at a fairly steady temperature while the cloud tops cool rapidly, encouraging the atmosphere to overturn and creating a thunderstorm peak at around two o'clock in the morning. Over the Great Plains, thunderstorms that may have originated above the slopes of the Rocky Mountains move eastward during the afternoon, feeding on the moisture in stiff southerly breezes ahead of them. These thunderstorms often arrive in places like Wichita and Kansas City after midnight.

The role of the Rockies in forming the Plains thunderstorms illustrates a basic fact about convection. Even an unstable atmosphere needs a trigger to start convection: Up-slope flow, air-convergence zones, passing fronts, and local heat and moisture sources all create bubbles with a difference.

Around the world, local convergence zones trigger thunderstorms regularly in the same spot. One such zone on a grand scale—the Intertropical Convergence Zone—is a shifting but permanent belt of thunderstorms near the Equator where easterly trade winds from the

At times hovering some 60,000 feet in the sky, the anvil-shaped cloud of a supercell thunderstorm caps a powerful storm system. The icy shield, flattened by upper-level winds, can spread far ahead of the massive updraft that produces it, casting an enormous shadow that often chokes off the development of storms far below.

hemispheres meet. The Indian monsoon is a regular feature in which moist oceanic air converges over a hot subcontinent from May to September. In the U.S., Florida readily fires up thunderstorms as sea breezes converge inland. The state is the country's thunderstorm capital, with many locations having more than a hundred storms a year. Similarly, lake and land breezes north of Lake Victoria in Africa yield a world-leading 242 thunderstorm days a year in Kampala, Uganda. West of the lake, Mbarara averages only 7 thunderstorm days a year.

Sometimes the thunderstorms get trapped along mountain slopes by a larger weather pattern. Runoff from heavy rain then funnels into narrow valleys, causing flash floods, the most prolific killer in a storm's arsenal. Up-slope winds along the Front Range of the Colorado Rockies were trapped this way in 1976, triggering an evening storm that unleashed a torrent of rain in Big Thompson Canyon. The simple convection cells turned out to be very deadly.

THUNDERSTORMS / FLASH FLOODS

On July 31, 1976, Colorado was one day away from its centennial. The people of the Mile-High State planned to celebrate their greatest resource, the majestic Rockies: A hundred teams of hikers prepared to race up Pikes Peak. Hundreds of thousands of others planned to enjoy the views afforded by the rugged spine of mountains dominating the American West. But that night, thunderstorms towered above the lofty crests surrounding Big Thompson Canyon. Weather and topography conspired to kill 139 people in raging water.

The flood came with startling swiftness. Campers at Sylvan Dale Guest Ranch had only a few minutes' warning before the waters rose. Five people drove away immediately and entered shallow water. Their car was swept off the road, capsizing in the black waters. Finally,

one woman was able to roll down her window and help the others climb out. The racing current carried her a quarter of a mile before she could grab a tree and hang on.

Twenty people made it to safety by abandoning their cars. Amid the roar of tumbling water, rock, and mud, they formed a human chain to help one another up the roadside and into a cabin. The waters demolished the road beneath them and lapped against the cabin, but the structure held firm, unlike 400 homes destroyed that night.

When the flood subsided, motels, campsites, and shops were gone. The canyon looked like a prehistoric landscape of boulders, logs, and mud. Such is the efficiency of a flash flood, the deadliest and most destructive force a thunderstorm can unleash. The strength of flash

Floodwaters shoved these cars off roadways and piled one atop the other near Vaison-la-Romaine after drenching rains in 1992 filled rivers to overflowing in southern France. The storms and their aftermath claimed 38 lives. When thunderstorms roll in and suddenly unleash their moisture, rain-splattered escape routes can wash away in treacherous torrents in a matter of mere seconds. As in the Big Thompson flood, which raced down a narrow Colorado canyon in 1976 and killed more than a hundred people, most flash floods in the United States trap victims in their vehicles, sweeping them away in fast-moving water as little as a foot or two deep.

floods is startling, and so is the ease with which storms can start them.

A thunderstorm over Unionville, Maryland, dropped more than an inch of rain in a minute. Another in Holt, Missouri, produced more than a foot in just 42 minutes. But circumstances, not storm severity, often trigger a flash flood: If the ground is saturated from previous rains, little rainwater is absorbed before reaching streams. In addition, the longer it rains, the slower topsoil absorbs water. The shape of terrain is also important, as is the timing of river crests. The flood crest in one major tributary reached Big Thompson Creek after the main, 30-foot crest had passed. Had the crests coincided, more people would have died.

While floods strike quickly, they often indicate slow-moving storms. A front may stall if it runs parallel to overhead steering winds or lies in a stagnant upper flow. Such a storm produced the steady rains that led to the 1889 Johnstown, Pennsylvania, flood, in which 2,200 people died. Similarly, weak upper-level winds allowed the Big Thompson storms to stall and release 14 inches of rain over a tiny area.

National Weather Service forecasters use radar that can quantify rainfall rates, signaling flash-flood potential. Research versions of the radar can also distinguish between hail and large raindrops, adding an important flash-flood detection capability. But a flash flood—by definition a sudden event—develops quickly, making it a formidable foe.

The Big Thompson flood was a sudden release of conditions brewing across Colorado and beyond. July 31 had been a beautiful day in Big Thompson Canyon, where U.S. 34 heads into the popular east gate of Rocky Mountain National Park. In most parts of the canyon, not a drop of rain fell, even during the flood. The day before, the Rockies had lain under unusually moist air pumped in at low levels from the Great Plains and at upper levels from the remains of daily storms over the Mexican Plateau. The air warmed during the afternoon of the 31st, boiling into thunderstorms mostly in Colorado's southwest corner, where upper-level divergence encouraged convection. But over the Big Thompson, in north-central Colorado, the quiet air near the surface needed a nudge upward to a level where it could flex its moist, convective potential.

The nudge came in part from a cold front—the edge of a massive dome of air spreading from the Dakotas. On the 31st the cold front advanced toward the foothills, a "back-door" maneuver in a region where storms mostly head eastward. The front brought in even more moist air that was ripe for making clouds. It gave a preview of its intentions in southeastern Colorado earlier that day, shooting off thunderstorms like fireworks as it lifted the warm, humid air ahead of it. By 5:30 p.m. the cold front jammed into the eastern foothills of the Rockies. Lifting by the terrain triggered the development of thunderstorms over the Big Thompson drainage basin.

Outflow developed under a thunderstorm near Denver, and the cold downdrafts spread northwestward at about 40 miles an hour. They kicked up a wall of dust, forming a gust front that merged with the strong flow into the foothills near the Big Thompson and accelerated the already rapid storm development. The turn of events shattered the pretense of stability masking trouble all afternoon.

National Weather Service radar in Limon, Colorado, sensed the explosion: At 5:30 p.m., the moist cold front arrived. A half hour later the gust front from Denver moved in, further lifting the moist air. The uplift triggered cumulonimbus towers up and down the foothills in Larimer and Boulder Counties. By 7 p.m. this line of convection was a ten-mile-high wall of boiling cloud. The ample moist air behind the cold front fed the thunderstorms. At upper levels, a weak southerly flow didn't hurry the storms. In fact, the towers were pinned against the mountains by strong low-level winds, just the way thunderstorms were wedged into the Black Hills one day in June 1972 when flash floods drowned 237 people in South Dakota.

Lots of moisture, pent-up instability, and light steering winds produced a classic scenario for heavy rains. The low-level air had only to rise 3,000 feet before saturating in low cloud bases. A few thousand feet more, and it reached free convection. Warmer than its surroundings, and moist enough to stay that way, nothing stopped the air from rising to nearly 60,000 feet in altitude, well past the top of the troposphere.

The storms just sat over the foothills and unloaded. Many thunderstorm rain shafts in Colorado meet dry air and evaporate before they hit the ground. But over the Big Thompson Canyon that day, the ice and water met more moist air, and the rain shafts proved frightfully efficient at reaching the ground. In four hours, more than six inches of rain fell east of the Continental Divide in a six-mile-wide area. Near Glen Haven, up to eight inches fell in less

than two hours. Over the Big Thompson Canyon near Estes Park, as well as over Storm Mountain to the north, totals reached ten inches.

The water tumbled down the slopes toward Big Thompson Creek with enormous speed. The solid, rocky slopes of the mountains absorbed little of the rainwater, contributing to the flood much the way miles of stone paving exacerbate urban floods—such as the infamous inundation of Florence, Italy, in 1966. Now the frenzied waters squeezed through the canyon toward Loveland, Colorado.

At 8:30, state patrolman Bob Miller checked out reports of rockslides. "The whole mountainside is gone," he radioed. "I'm going to get out of here before I drown." Patrolman Hugh Purdy went looking for the flood crest, but it found him. "I'm right in the middle of it. I can't get out...," he radioed only a half mile from Drake. Fifteen minutes later, flow in the Big Thompson near Drake had grown to more than 200 times its normal rate. The little mountain stream carried twice the flow the mighty Mississippi carries in Minneapolis. Water in the narrow canyon rose 19 feet in minutes, swamping U.S. 34. Purdy was swept eight miles downstream, his car mangled beyond recognition by the boulders and other debris tumbling in the water. The car's remains were identifiable only by a State Patrol key ring.

All but ten miles of U.S. 34 washed out. Survivors who fled to high ground had to share it with thousands of rattlesnakes that escaped the waters. The next day close to one thousand people were evacuated by helicopter from the Big Thompson Canyon. One helicopter crew spotted a cowboy hat on the silt and gravel left

TIM WILEY

Once a pleasant riverside retreat, this house came to rest some 200 yards from its foundation after a flash flood roared through northern Colorado's Big Thompson Canyon in 1976. Of some 5,000 flash floods that strike the United States each year, only a few achieve the total devastation that Big Thompson Creek, normally a few feet deep in places, wrought in several terrifying minutes. A massive storm, hovering for hours over the foothills of the Rocky Mountains, had dropped nearly a foot of rain into the narrow canyon.

by the flood, and took a closer look. A man was buried in mud up to his chest, lucky to have been standing when the waters hit him.

It had been the worst flood in the state's hundred years. In Denver, at a ceremony to cut centennial cake, the governor prayed for the victims of the thunderstorm's sudden fury.

THUNDERSTORMS / JOHNSTOWN 1889

"It seemed as if a forest was coming down upon us. There was a great wall of water roaring and grinding swiftly along, so thickly studded with the trees…that it looked like a gigantic avalanche."

So the Conemaugh River looked to a survivor of the disaster on May 31, 1889. After rising all day in steady rains, the river's waters leaped upward with the break of a dam above Johnstown, Pennsylvania. In an instant, a flood became one of the worst man-made tragedies in American history—a weather event turned disastrous by years of foolhardy neglect.

The Alleghenies of west-central Pennsylvania are productive mountains that hold iron ore, limestone, and coal. And Johnstown, at the flood-prone bottom of the Conemaugh River Valley, was an industrious town thriving on these resources. Unfortunately, the company town for Cambria Iron and Steel lay just 15 miles downriver from an artificial resource that was decidedly unsound. More than three decades before, South Fork Creek had been dammed for the Pennsylvania Canal.

In 1879, after the canal was abandoned, the reservoir at South Fork was turned into a sporting resort for wealthy executives and their families. The mud-and-shale dam was expanded and topped with a road. But over the next decade leaks appeared, and the dam owners did little more than patch the precipitous earthen wall with straw. Residents of the valley below complained to little avail. A blue-ribbon panel of engineers inspected the hundred-foot tall structure and determined that it was safe—as long as water did not flow over it.

The dam usually cleared the reservoir level by 15 feet, but the storms in May 1889 proved more than its match. Late in the month, a storm system moved in, bringing Canadian air across the border while warmer air moved west to meet it. On Memorial Day raindrops began to fall over Johnstown, and the next morning, Friday, the 31st of May, the rain continued. Residents watched the Conemaugh River rise 20 feet above its usual level, setting a record. Up to 9.8 inches of rain fell in nearby Wellsboro, but more probably fell in the Conemaugh basin. No one knows exactly how much water fell in Johnstown, because the weather observer there did not live to tell.

A few people heeded warnings about new leaks at South Fork Dam on the fateful Friday and left for high ground, but others in Johnstown, tired of such frequent worries, derided the cautious. Workmen were already busy reinforcing the dam that day when the water topped the earthen bulwark, eroding it slowly through the early afternoon. Rev. G. W. Brown of South Fork went to inspect the leak and saw a foot of water spilling over the dam. Then at 3 p.m., the structure's wood frame gave way, blasting trees and mud and helpless workers skyward, and opening a hole wide enough for a train to plow through. "God have mercy on the people below," the pastor exclaimed.

The breach spread to nearly 100 yards wide. One observer said the 70-foot-deep lake began showing bottom in only 5 minutes. In less than an hour the lake basin was drained. The muddy waters tumbled down the 450-foot drop toward 30,000 souls in Johnstown. On the way passengers in a train that had stopped on the water-covered tracks looked on in horror as the engine

Floating, combustible debris—carriages, walls, trees, and train cars— turned the raging Conemaugh River of west-central Pennsylvania into a deathtrap for Johnstown residents in 1889 (above). Most people had ignored signs of imminent collapse by South Fork Dam, an earthen barrier holding back rising waters in a reservoir upriver. One survivor later posed for a picture atop an uprooted, house-piercing tree (right).

was wrenched into a whirlpool. The already flooded Conemaugh doubled when a wall of water 35 feet high tumbled downslope at nearly 16 miles an hour.

People in the valley heard the approaching avalanche of water. A blast of moist air snapped tree branches, and from a distance the water looked like a cloud of black smoke. Some thought a fire or dust storm was racing toward them. The crest demolished South Fork, a town of 2,000 people, and Mineral Point, a settlement of 800. At Conemaugh, six miles downstream, people heard the rumble of the approaching torrents, but assumed it was thunder. Then the flood hit, its waters carrying houses and bodies from the towns above. Locomotives in the East Conemaugh railroad yards were buried in mud.

At 3 p.m. the frightened Western Union telegraph operator in Johnstown wired the Pittsburgh office to say she would have to abandon her post. She had already moved to the second floor of the office when the river began inundating the building; now the river was rising up to her knees again. Seven minutes later the line went dead.

The company-owned tenements in the flats along the river were scoured from their foundations and smashed as a mass of lumber, boulders, and bodies pressed against the stone arch bridge spanning the river in Johnstown. Hundreds of people fortunate enough to escape the deluge clambered onto the old bridge. But then the debris ignited, engulfing them in floating flames.

After the flood more than 30 acres of debris choked the valley. Looters and tourists clambered over the ruins of Johnstown, marveling at trees that had pierced roofs like battering rams. More than 2,200 people had paid for a weak dam with their lives.

THUNDERSTORMS / LIGHTNING

Lightning was the weapon of choice for Zeus and Thor, two of the supreme gods of the ancients. Though today the towering cumulonimbus does not rise to the rank of deity, lightning still defines it. With lightning, a thunderstorm strikes at will and kills not just Homeric heroes like Ajax, who offended Athena, but nearly a hundred Americans every year. It remains the second-deadliest force of weather in the United States, ahead of tornadoes and hurricanes.

Mythic force though it is, lightning energy is small change in the vast riches of heat and buoyancy expended by a thunderstorm. The number of lightning bolts in a storm is not a reliable indicator of its strength. These sizzling tantrums heat air to six times the sun's temperature, but meteorologists can easily omit them from their computer models of storms. In their excess of updrafts and ice, thunderstorms have an effortless flair for making electricity fly.

When positive and negative charges are separated, an electrical field is established. Lightning is the sudden relaxation of a very strong field—a high-speed delivery of millions of trillions of electrons. Electrification can occur anywhere that particles collide: in the plume of a volcanic eruption or in the fireball of a thermonuclear explosion. Lightning is found in snowstorms and dust storms and even on other planets. Thunderstorms are particularly good at banging airborne particles together, just like hydrogen bombs and volcanoes. But scientists are still fleshing out the complicated means by which charges are separated in storms.

Benjamin Franklin first definitively established the electrical nature of storms in 1752.

With 100,000 thunderstorms annually nationwide, most of the United States gets more than 10 lightning strikes per square mile each year. Even more bolts of lightning grace the clear desert vistas of thunderstorm-prone Tucson, Arizona (opposite), where high-voltage spectacles staged by nature overpower bright lights in the city below.

Franklin observed that most thunderstorms are negatively charged. For years, scientists depicted thunderstorms as a dipole, negative at the bottom and positive at the top. Actually, aside from a very thin layer of negative charge at the top, a thunderstorm is more like a tripole. The main region of positive charge is in the storm's upper reaches, but a lesser region of positive charge is in the cloud base; in between is a mile-deep concentration of negative charge about four miles from the earth's surface.

The middle negative layer resides where the cloud is about 5°F, a temperature at which all phases of water can coexist. Supercooled liquid water (below 32°F) helps charge separation in two main ways. First, it fuses to ice crystals and forms graupel, the tiny hail embryos high in the storm. Second, its presence allows graupel to acquire a significant charge when it collides with lighter ice crystals. Interestingly, at about 5°F, the charge separation in graupel-crystal collision shifts: At lower temperatures, the graupel becomes negative; at higher temperatures, it becomes positive. Thus the storm's workhorse updrafts not only form rain and hail, they separate charges. The updrafts produce ample collisions of graupel and ice crystals—meteors of different size and weight—and bring light water droplets high above the 5°F level, enabling graupel to form and charge. They also

sort charges after collisions: Negative graupel stays in the middle of the cloud while smaller, positive ice crystals are swept higher.

Scientists at the National Center for Atmospheric Research in Boulder, Colorado, soar into thunderstorms in a glider, riding the thermals and measuring electrical properties. They have found that the electric field in a thunderstorm can triple in a minute. In a few minutes, the updrafts create an electric field strong enough to create a spark. The air between the separated charges is an effective insulation only to a point, because the electric field eventually becomes so strong that a lightning channel can

form. With large voltage differences between cloud layers, most lightning stays within the storm cloud, rather than striking the ground or (rarely) another cloud.

Cloud-to-ground lightning usually begins from the cloud, near the base. A spark invisible to the eye races downward on a jagged path of approximately 50-yard-long steps, creating a channel of air molecules stripped of electrons. When this spark, called a stepped leader, gets within perhaps 100 yards of a grounded object, a responding spark surges up from the object. Upon connection, a luminous stroke of lightning zaps up the completed channel at about

60,000 miles a second. Thus, in slow motion, lightning looks like it propagates upward, even though the channel was initiated from above. And in fact, electrons flow downward—but those closest to the ground respond to the connection first. The luminosity progresses upward as successive electrons farther up the channel respond to the connection.

The established connection may experience multiple strokes within a fraction of a second, all adding up to one lightning flash. The average flash has three or four strokes with about four-hundredths of a second in between. But some flashes have 20 or 30. The multiple strokes cause the light to flicker in the channel, but, like individual frames of a movie, these are not always visible: The stroke (but not the whole flash) is quicker than the eye.

About one hundred lightning bolts strike earth every second, making lightning a significant threat. Airplanes and skyscrapers have brought us close to the clouds, and both get struck by lightning frequently. A plane circling for landing in Chicago was struck four times in 20 minutes. The Empire State Building gets struck by lightning an average of 23 times a year. In one storm it got hit 8 times in 24 minutes. The Bell Tower of San Marco in Venice has been destroyed three times by lightning; after a lightning rod was installed in 1766, the tower has not suffered damage. Unfortunately, until that time Europeans commonly believed that lightning could be prevented by ringing local church bells at the sign of a growing thunderstorm. In one 33-year stretch, more than 100 bell ringers died from lightning. Long after the days of bell ringing, scientists have shown that tall

LIGHTNING AND THUNDER

■ With lightning a thunderstorm both dazzles the eye and shakes the ear. The sudden heating of a lightning channel explodes air outward in a blast that creates a long-lived sound wave—thunder.

■ Sound waves travel at only 760 miles per hour and weaken with distance in the atmosphere. This creates a variety of effects. Thunder can begin with a bang of sound from near parts of a channel, then roll as sound arrives from farther away. A flash angling toward you might roll off into silence slowly; in a flash that begins overhead yet attaches to ground far away, the thunder from up and down the channel may reach your ears simultaneously in a short, crumpling sound.

■ Thunder is usually inaudible beyond about 15 miles, because the decrease of temperature with height in the atmosphere bends the sound waves upward away from your ears. As a result, a distant cloud top may flash soundlessly—some people misname it heat lightning—simply because the thunder passes far overhead.

■ Unlike sound, light reaches you nearly instantly. Every five seconds between the flash and its thunder represents about a mile between you and the closest part of the lightning channel. Of course, no counting will tell you where lightning will strike next.

■ And long after the rumbling has stopped, the rain has ceased, and the sky has begun to clear, lightning can strike. These seemingly inexplicable bolts from the blue sometimes come from the backside of thunderstorms far over the horizon; a lightning bolt can follow its tortuous path for dozens of miles.

buildings actually initiate a stepped leader upward, drawing a response from the cloud in a reverse lightning stroke.

Temperatures in a lightning channel can reach 54,000°F. Fortunately the channel is only about an inch wide. This tiny channel can easily destroy a tree: When lightning strikes a tree, the wood resists the sudden surge in electricity,

Lightning surges through the body of a tree while a thunderstorm rages. Frequent targets of lightning, trees can put people standing beneath them right in the line of fire. The briefly energized lightning channel—four times the temperature of the sun's surface and twenty times the pressure of the surrounding air—can vaporize and explode trees and other objects that resist its awesome current.

causing the temperature along the channel to rise enough to vaporize the wood. If the channel is in the core of the tree, the trunk may explode. Lightning triggers about 10,000 forest fires a year in the U.S., and as many as one half of lightning discharges—the ones with currents lasting hundredths of a second, not just millionths—can spark a tree to flames. Adding to the fire danger are dry thunderstorms typical in the American West, where lightning sparks fires without the attending rain to put them out.

Much research today investigates what happens when lightning carries positive, not negative, charge. Such positive discharges come from a thunderstorm's anvil or other positively charged regions. These bolts are powerful, if infrequent, and carry the sustained current that starts forest fires. They usually transfer to the ground three or more times as much charge as negative flashes—but in one stroke.

Lightning makes thunderstorms the most spectacular part of a never-ending electrical dialogue between the earth and its atmosphere. Overall, the surface of the earth is a negative link in the global electric circuit, and the atmosphere—especially 30 and more miles high—is positive. Thunderstorms are the batteries that continually recharge the earth. In addition, a thunderstorm produces a "shadow" of positive charge on earth's surface. Sometimes the electrical field from a storm is too weak to initiate lightning but strong enough to cause St. Elmo's Fire, a corona of positive charge emitted from sharp objects like aircraft wings and ship masts.

In recent years scientists confirmed some bizarre skyward emissions from thunderstorms. For years, night pilots claimed they saw blue flashes above thunderstorms, but no one could photograph these brief and seemingly rare displays. Then, black-and-white movies taken from the space shuttle recorded several upward lightning bolts reaching high into the stratosphere. Today, from airplane windows and mountaintops, researchers film these discharges, documenting a whole new class of lightning-like phenomena—red sprites, elves, blue jets, and more—that continue to expand the lore of thunderstorms.

THUNDERSTORMS / HAIL

When a thunderstorm drops heavy objects, they are often encased in ice. A chunk of ice eight inches across dropped into Bovina, Mississippi, in 1894. Inside was a frozen gopher turtle. In 1930, five glider pilots bailed out in a storm over Germany. Carried into the dense moisture of the freezing cloud interior, they were covered by ice. Miraculously, one of them survived the fall.

Even when the heavy meteor is simply ice—hail, that is—it makes a remarkable sight indeed. Certainly hail was as impressive as any of the plagues that struck biblical Egypt. These icy deluges are evidence that awesome powers are at work overhead—the powers of updrafts.

Air in a strong thunderstorm easily rises fast enough to support heavy objects. It is nearly impossible for a person to stand in a 70-mile-an-hour wind. Imagine the effects of updrafts clocked at 100 miles an hour. Water rises with this rushing air high into the subfreezing reaches of the atmosphere. Sometimes air rises so fast that large droplets or ice crystals do not form in the core of the updraft; they form near its edges. This leads to a telltale radar signature of hail formation: a curtain of high reflectivity adjacent to a vault of radar-transparent updraft.

Near the edges of the updraft's core and between 15,000 and 30,000 feet, ice, water, and vapor are plentiful. A typical accumulation region might hold more than half a pound of water for every cubic foot of air. If an updraft hits 70 miles an hour, and there is enough water in the cloud, the hailstone can grow to 3 inches before hitting the ground.

When hail becomes too heavy for the updraft, it can be deadly. In April 1888, hail killed 246 people in northern India. A two-hour ice deluge in China's Hunan province killed 200 others in 1932. In Alberta, Canada, in July 1953, two hailstorms four days apart killed 64,000 ducks and thousands of other wild birds.

Hail often grows spiky protrusions. Yet, for all its irregularities, it actually grows in an orderly fashion. When scientists cut into a hailstone, they see growth rings. At the center is a hail embryo, which often starts as an ice crystal or a frozen water droplet. Liquid water droplets supercooled below their freezing point fuse onto the embryo, and a thin frozen layer builds a small crystal called graupel. The graupel continues to accrete layers, reaching hailstone status when its diameter exceeds a fifth of an inch.

The layers of the completed stone are clear or frosty, depending on how quickly the supercooled droplets freeze. Clear layers accrete in warmer regions of the cloud with abundant liquid water. There the water takes more time to freeze, thus eliminating air bubbles; frosty layers come from cold, drier regions where water froze too fast for trapped air to escape.

Strong thunderstorms generally have a greater potential for big hail. A Coffeyville, Kansas, storm on September 3, 1970, produced a stone 17.5 inches in circumference and weighing 1.67 pounds. A 4.2-pound hailstone fell in Kazakstan, and a 2.1-pound stone landed in Strasbourg, France, in 1958.

The strength of a thunderstorm does not ensure that hail will hit the ground, however. A 10,000-foot-thick layer of air at above-freezing temperatures is usually enough to melt large hailstones into big raindrops. This is one reason why Florida and most tropical lands have little

Ice falling at close to 100 miles an hour pocked automobile windows in Tampa, Florida, during a 1992 thunderstorm that cost some 25 million dollars in property damage. Although hailstones can come in uniform, well-rounded shapes, they often fall as irregular ice chunks, including a softball-size missile collected in West Texas (right).

hail. But where land rises to meet the clouds, hail is quite frequent. The tropical highlands near Africa's Lake Victoria have more hail than anywhere else: Kericho, in the Nandi Hills of Kenya, averages 132 hail days a year. The Andes highlands, southern France, northern Italy, Romania, and the Caucasus can average 10 hailstorms a year. Nearly 5,000 hailstorms strike the United States each year, many in the Great Plains, where hail frequency ranges up to about 10 storms a year. Some locations in the Rockies may get 20 hailstorms a year.

These storms produce more widespread damage than tornadoes, which often come from thunderstorms with stronger, more concentrated updrafts. A storm in Alberta, Canada, once created a 20-mile-wide, 200-mile-long swath of walnut-size hail. An Australian storm in 1913 reportedly covered 17,000 square miles with

hail. A storm in Illinois in 1968 took 90 minutes to drop an estimated 82 million cubic feet of ice on a 19-mile-wide, 51-mile-long swath.

Such hail swaths long have been a scourge of agriculture. Devastating hail in 1788 led to a famine that helped precipitate the French Revolution. Today hail still causes hundreds of millions of dollars in crop damage each year, but it wreaks devastation on cities as well. The most expensive thunderstorm in U.S. history hammered Fort Worth, Texas, with hail on May 5, 1995. People who had gathered at an outdoor festival ran for their cars when spiked, softball-size ice tumbled out of the clouds at a hundred miles an hour. Miraculously no one was killed by the hail, which injured a hundred people and caused two billion dollars in damage.

The cost of such modern deluges in cities of glassy high-rises and vulnerable car windows mounts in an age when the mystique of ready-made ice in summer fades. Imagine the delight in Union County, Iowa, in 1890 after a thunderstorm dropped four inches of hail. Piled by wind into six-foot drifts, the ice took a month to melt. One local scooped up some for himself and made a gallon of ice cream.

THUNDERSTORMS / MICROBURSTS

The powerful updraft was once the star of a thunderstorm's show, with electrical wizardry, damaging hail, and flooding rains. Downdrafts attracted little attention—few scientists believed rain-cooled air sinking out of a storm posed a significant threat. Then a string of airplane accidents killed more than 500 people between 1970 and 1985. Each tragedy was linked to a thunderstorm. And one led to a meteorological breakthrough.

On June 24, 1975, rain from a thunderstorm shrouded the approach to New York's Kennedy Airport. An Eastern Airlines jet crashed, killing all but 11 of 124 people on board. An airline executive, suspecting wind shear, asked T. T. Fujita of the University of Chicago to investigate.

Near the end of the runway, Fujita found evidence of three small, strong downdrafts fanning out surface winds at 55 miles an hour. Oddly, 11 planes had descended successfully through the rainy downdrafts shortly before the crash. The only planes to encounter the high-speed horizontal wind were the three that passed through the downdrafts below 700 feet. Fujita dubbed the killer combination a downburst—a downdraft with a low-level outburst.

A downburst is a pulse of dense air in the downdraft of a mature thunderstorm. This pulse can head earthward at about 40 miles an hour, and when it hits land, it can go nowhere but out. A microburst is a downburst less than 2.5 miles wide. These local winds are hard to monitor. A microburst near Andrews Air Force Base descended only minutes after President Reagan landed in Air Force One. Toppled trees indicated 130-mph winds, yet nearby wind gauges never registered anything stronger than

RONALD D. BREWER

11 miles an hour. The discovery of powerful microbursts opened meteorologists' eyes to a new class of storm damage. Shortly after his analysis at Kennedy Airport, Fujita showed that suspected tornado damage in a Wisconsin town was, in fact, caused by microbursts. Finally, sailors and wildfire fighters also had an explanation for the sudden thunderstorm winds that bedevil them. Research showed that Chicago could average more than one microburst a day in springtime; near Denver, another study revealed 186 microbursts in less than 3 months.

The compactness of a microburst is a special threat to pilots, who cannot always compensate for sudden changes in wind speed and direction. The wind shift in a microburst can top 100 miles an hour, but even an average shift, close to half that, will cause trouble. In the New Orleans area in 1982, a Pan American jet headed into a 17-knot wind at takeoff, but within seconds

THIERRY CAMPION/SYGMA

Embedded in a thunderstorm's rain shaft near Rutland, Illinois, a microburst reveals itself by spinning up dust in outflow near the ground (opposite). Fleeting wind shear, which can bring down a jetliner, probably caused the 1985 crash of Delta 191 (left), lying in charred remains short of a Dallas-Fort Worth runway. The fiery disaster took the lives of 137 people.

sagged under a downdraft, then lost lift in a 31-knot tailwind. More than 150 people died.

At first, some of Fujita's colleagues were skeptical that downdrafts could create the strong near-surface winds aircraft were encountering. But researchers showed that an outburst formed a ring-like vortex that intensified the wind shear. After hitting the ground, the leading edge of the outflow rolls over, and the foot of the storm's rain shaft visibly curls its toes upward. This vortex ring on the ground expands outward and lasts up to 30 minutes.

The stronger the downburst, the more rings it might create in pulses minutes apart. A Delta Airlines jet descending at Dallas-Fort Worth in 1985 may have encountered multiple downburst rings. The crew maneuvered through a gain and loss in airspeed as it passed through a downdraft, only to crash a mile short of the runway in another loss of airspeed moments later.

Not all microbursts come with intense rain. Near Denver, storms may have bases 10,000 feet above the High Plains. Precipitation evaporates before reaching the ground, but the downdraft it initiates can cause a "dry" microburst. A telltale dust ring may spread out below the storm. Today major airports have Doppler radar and other instruments that detect the surface winds of microbursts. Still, a microburst may have claimed a jet in Charlotte, North Carolina, as recently as 1994.

Ironically, our understanding of such storm dangers stems from a deadlier event in history. Trying to imagine what small-scale thunderstorm wind could down a jet, Fujita examined a starburst of damage he had photographed after a West Virginia tornado in 1974. He recognized the pattern as a sign of blasting winds in part because it matched one he had surveyed early in his career in Japan, in the rubble of Nagasaki.

THUNDERSTORMS / DERECHOS

When thunderstorms band together, watch out. Early on the morning of July 15, 1995, forecasters in upstate New York could hardly believe their eyes. Satellite and radar readouts showed a line of thunderstorms sweeping across Lake Erie at 80 miles an hour. Lifting unusually hot, humid air in the Adirondacks, the storm line ignited new convection as it raced toward the Atlantic; 3,000 bolts of lightning a minute laced the night sky.

The storm line accelerated to 85 miles an hour, with 106-mile-an-hour gusts in Watertown. Fronts often produce strong gusts, but these winds were sustained at more than 50 miles an hour for over 20 minutes. Millions of trees in Adirondack State Park were uprooted or snapped in two.

A wide, long swath of damaging winds is called a *derecho*, which is the Spanish word for "straight" (in contrast to tornado, which comes from the Latin word for "turn"). This wind is usually caused by a group of thunderstorms that creates a mini-weather system with a broad plow of outflow 300 miles wide. Derechos, which cause a path of damaging winds 250 miles long, are far larger than tornadoes and, in fact, can spawn tornadoes of their own. They are prime examples of what happens when thunderstorms reinforce each other.

When a thunderstorm's rain-cooled downdraft hits the ground, it can spread out a small gust front, kicking up dust and lowering temperatures locally. The gust front noses into warm, moist air ahead of the storm, forcing it upward and sometimes initiating new convection. When several nearby storms do the same thing, the leading edge of cooled outflow surges ahead of the storms and forms a much wider gust front. A derecho is a very wide, long-lived gust front that lifts air in its path like a plow, thus helping to perpetuate the original group of storms. A July 1993 derecho lasted 12 hours, tracking 500 miles from Colorado and hitting a 70-mph stride as its 125-mph gusts tossed trucks off Interstate 80 in Nebraska, uprooted trees, and caused more than 100 million dollars in damage. Cool air finally stopped the self-fueling wind in Iowa.

In 1980, scientists identified a persistent type of self-perpetuating thunderstorm alliance now carrying the unwieldy name of mesoscale convective complex (MCC). One system can cover a state, making it smaller than the large frontal systems that place big "L"s on weather maps, but much larger than the low-pressure area under an individual thunderstorm. Between March and September, up to 80 of them roam the United States east of the Rockies.

Most MCCs tend to last 12 to 24 hours and prowl the Plains and Ohio Valley after sunset, but the same MCC might disappear for the day, only to rise again at evening, night after night. This mysterious cycle starts when thunderstorms begin to warm the air around them. Air from miles away starts to flow toward this buoyant region at mid and upper levels of the clouds, turning the MCC into a large area of warm, moist air efficient at producing rain. Meanwhile, the outflow at the tops of the central storm towers spreads out in a unified anti-cyclonic circulation, blocking upper-air streams in the area. Winds approaching the MCC must detour around it, a divergence aloft that stokes the flames of convection within the MCC. By

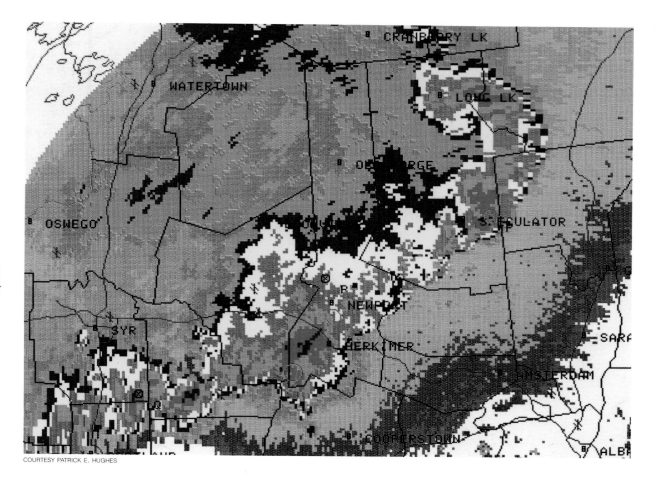

Exploding to 70,000 feet in height, the towering storm line of July 15, 1995, first showed up on radar screens while still hundreds of miles beyond the horizon. The resulting derecho packed its strongest winds at the curling northeast end of the line of storms (yellow and red), shown here at 4:51 a.m., battering forests east of Watertown, New York.

maturity, the circulation creates a well-rounded mass of convection—thunderstorms circling their wagons in the prairie night.

Nighttime conditions over the Plains and Southeast reward these storms. A fast flow of moist air at a couple of thousand feet forms regularly at night. This air supply from the Gulf of Mexico is similar to low-level moist flows over the Atlantic Seaboard, as well as coastal China, Argentina, and Australia. Called a low-level jet, the wind helps make up for the loss of solar heating after sunset, fueling the MCCs at night. As a result, most MCCs peak when the southerly flow peaks, around midnight.

The MCCs help sustain the nation's breadbasket. Moving slowly at about 20 miles an hour, an average MCC spreads about 800,000 tons of rain over an area twice as large as the state of Iowa. Up to 70 percent of spring and summer rains for the farms between the Rockies and Mississippi River can come from MCCs and similar, but less powerful, thunderstorm clusters.

These nocturnal storm systems cause heavy rain, hail, and tornadoes when people are most vulnerable—a quarter of MCCs claim casualties, many in flash floods. An MCC brought flash floods back to Johnstown, Pennsylvania, on July 20, 1977, when more than 8 inches of rain fell in 7 hours, killing 77 people.

A thunderstorm cluster can begin as a squall line—a row of storms perhaps a hundred miles long parallel to, and racing ahead of, a cold front. At times, rain falls from the relatively flat clouds stretching behind the squall-line storms. But by giving storms a chance to help one another, squall lines can cause some of the most damaging thunderstorms in the world. In April 1974, a well-developed storm system crossed the Mississippi River and produced 3 of the most violent squall lines in history, spawning 148 tornadoes in less than 24 hours.

THUNDERSTORMS / TORNADO OUTBREAK

Amazing things happened on April 3, 1974. Some people saw flying two-by-fours pierce their houses. Some saw green and pink and black-yellow clouds hurl baseball-size hail. Others heard what they thought were freight trains overhead. "I saw a big, two-story house comin' over the hill," another said.

One woman, alerted by a friend's phone call, went to the window and saw a tornado approach. She dived under a pool table and survived. The chair where she had sat seconds before was crushed under the fallen chimney. Another woman pulled and pulled but could not get the storm-cellar door to shut as winds began to scream overhead. She looked up and saw half the house had blown away.

In less than 20 hours, beginning at 1 p.m. central time on April 3, 148 tornadoes struck the United States east of the Mississippi River, killing 315 people and injuring more than 5,000. Nearly 10,000 homes were destroyed and 15,000 damaged. The largest tornado outbreak in history scarred 13 states and one province from Mississippi to Ontario, Canada. Lined end to end, the tracks of the tornadoes stretched 2,500 miles—in one day nearly equaling the combined swaths of all of the twisters of 1972.

Outbreaks leave their mark nearly every year. The 51 tornadoes from Iowa to Michigan on Palm Sunday 1965 killed 256 people. In June 1992, a month that averaged a dozen tornadoes a day, 123 tornadoes struck on the 15th and 16th from Kansas to Minnesota, and an outbreak of similar proportions riddled the Midwest and South in April 1996. But none of these days could match the Jumbo Outbreak, when the

atmosphere primed for violence let loose on a historic scale.

Forecasters saw tornado potential for that Wednesday more than a day ahead. Already a 20-tornado outbreak had stung the area on Monday, and another low-pressure area was approaching from the Rockies with cold air north and west of it and warm, moist air east and south of it. The circulation covered more than a million square miles.

Overnight, typical severe thunderstorm conditions developed: A tongue of moist Gulf air surged into the Midwest and Ohio Valley ahead of the strong surface cyclone. Dry southwesterly air slipped in atop the surface moisture. Meanwhile sinking air aloft clamped down on the steaming, inverted brew. Ready to rise at the slightest stirring, the brew cooked while the low approached from the southwest.

The low moved north on Wednesday under the guidance of the jet stream, which turned sharply northeastward from Texas through Oklahoma and Missouri into Illinois. At 1 p.m., the central surface low had reached the Illinois-Iowa border, spreading snow across the northern Plains and into northern Michigan. Meanwhile, the strong southerly flow high above proved friendly to the storm, which organized into a complicated arrangement of two cold fronts and two warm fronts.

Ahead of one cold front was a sharp moisture boundary, called the dry line, running north through Louisiana to Illinois. These boundaries, and strong divergence around the jet-stream flow, stirred the warm surface brew in the eastern states. Three squall lines broke out in the warm sector of the cyclone as it

As severe thunderstorms go, so go tornadoes. Tornado tracks from the Jumbo Outbreak of 1974 largely paralleled the environment steering the thunderstorms. Exceeding 120 miles an hour, northeastward winds above 15,000 feet pushed the storms ahead a mile a minute. In all, the parent cyclonic system covered a million square miles of the United States, though the tracks themselves averaged 16 miles in length.

IOWA

MICHIGAN

NEW YORK

Detroit

Chicago

Morris

Toledo

Cleveland

Fort Wayne

PENNSYLVANIA

ILLINOIS

Pittsburgh

INDIANA

OHIO

Decatur

Columbus

Indianapolis

Dayton

Xenia

MD.

Cincinnati

St. Louis

Lexington

Madison

WEST VIRGINIA

Louisville

Lexington

Brandenburg

Beckley

VIRGINIA

MISSOURI

Roanoke

KENTUCKY

Corbin

Nashville

Knoxville

NORTH CAROLINA

ARKANSAS

TENNESSEE

Asheville

Chattanooga

Blue Ridge

Huntsville

Columbia

Atlanta

SOUTH CAROLINA

Birmingham

GEORGIA

Tuscaloosa

Macon

MISSISSIPPI

ALABAMA

Montgomery

Tornado Path

Laurel

0   100 mi
0   100 km

spread. Many thunderstorms in the warm sector of a sprawling extratropical system will fire up over a local hot spot and then boil until they exhaust the warm fuel below. But squall lines, pushed along by upper-level currents, can gobble up fuel over a vast area. The faster eastward they move the more moisture they ingest. In the Jumbo Outbreak, the squall lines moved at 55 miles an hour. The rows of thunderstorms ballooned to dizzying altitudes, some reaching 65,000 feet. Every tornado was planted by one of these lines of convergence that plowed up the warm, moist fuel hugging the ground.

Before any of the twisters arrived, baseball-size hail injured 25 people in St. Louis. Then, at 1:10 p.m., a weak tornado scouted Morris, Illinois, for only a minute. After a 50-minute lull, the storm began launching tornadoes in earnest. Nearly simultaneous tornado strikes in Illinois, Indiana, Tennessee, and Georgia opened the main barrage: 93 twisters in 8 states in only 7 hours. The fast upper-level winds propelled many of the tornadoes at 60 miles an hour, most tracking northeastward.

In northern Alabama a killer twister laid a 51-mile track of terror, wiping away houses. After crossing Wheeler Lake as a monstrous waterspout, the 260-mph winds sideswiped mobile homes. One injured man was taken to a church less than a mile away to wait for an ambulance. A half hour later he was killed when a second tornado plowed into the church.

The most devastating tornado struck at 4:32 near Xenia, Ohio. Winds topped 260 miles an hour, destroying or damaging the homes of nearly half the town's 25,000 residents, killing 34 people and injuring another 1,150.

Xenia could thank daylight savings time for averting an even grimmer tragedy: Six schools were badly damaged—three of them totally destroyed—shortly after most students had left. Three school buses were hurled onto the stage of the high school's auditorium just after drama students had fled to the hallway. The roof collapsed onto the seats behind them.

One student, seeing a tornado approach his school bus, convinced the driver to stop and evacuate the vehicle in the nick of time. The bus blew over as the students lay in a ditch. Not all children were so fortunate. In Lexington, Indiana, a dozen other children were injured when their school bus was toppled by a tornado. Another bus was blown 400 feet off the road, injuring the driver and killing his wife.

Two brothers driving on a country road in Madison, Indiana, pulled over to the shoulder of the road when hail began to dent the car. After ten minutes, they continued on their way, amazed at the damage around them. The hail had been so bad they hadn't realized they had survived the core of a mile-wide tornado.

As the winds picked up, a woman in a Louisville, Kentucky, record store was blown across the shop as a shattered wall of glass sprayed across the merchandise and shoppers. A Xenia family was eating an early dinner when the father got up to find out why the back door was banging. He looked out and saw his neighbor's house disappear in the whirlwind. The family dived to the floor and survived as the roof lifted from their house and the walls collapsed around them.

One man drove home from work in the ominous weather. When he turned onto his street

THE FUJITA SCALE

■ **F0: Gale tornado (40-72 mph).** Some damage to chimneys; breaks branches off trees; pushes over shallow-rooted trees; damages signboards.

■ **F1: Moderate tornado (73-112 mph).** The lower limit is the beginning of hurricane wind speed; peels surface off roofs; mobile homes pushed off foundations or overturned; moving autos pushed off the roads; attached garages may be destroyed.

■ **F2: Significant tornado (113-157 mph).** Roofs torn off frame houses; mobile homes demolished; boxcars pushed over; large trees snapped or uprooted; light-object missiles generated.

■ **F3: Severe tornado (158-206 mph).** Roof and some walls torn off well-constructed houses; trains overturned; most trees in forest uprooted; heavy cars lifted off ground and thrown.

■ **F4: Devastating tornado (207-260 mph).** Well-constructed houses leveled; structures with weak foundations blown off some distance; cars thrown and large missiles generated.

■ **F5: Incredible tornado (261-318 mph).** Strong frame houses lifted off foundations and carried considerable distances to disintegrate; automobile-size missiles fly through the air in excess of 110 yards; trees debarked; steel-reinforced concrete structures badly damaged.

■ **F6: Inconceivable tornado (319-379 mph).** These winds are very unlikely. The small area of damage they might produce would probably not be recognizable along with the mess produced by F4 and F5 wind that would surround the F6 winds. Missiles, such as cars and refrigerators, would do serious secondary damage that could not be directly identified as F6 damage. If this level is ever achieved, evidence for it might only be found in some manner of ground-swirl pattern, for it may never be identifiable through engineering studies.

he saw a tornado approach. The funnel picked him up; his Volkswagen somersaulted twice, landing in a neighbor's yard. The man, badly cut by flying glass, rushed into the ruins of his house to find his family safe in the basement.

Individual twisters that day were as amazing as the scale of the outbreak. Tornado expert T. T. Fujita of the University of Chicago flew more than 10,000 miles to survey the area. The photos were a "gold mine" of severe weather oddities. Fields showed overlapping cycloidal marks indicating multiple vortices circling within larger funnels. In one case, two tornadoes had swung around each other as if in a country square dance. One tornado near Blue Ridge, Georgia, hiked up one slope of Betty Mountain, a 3,300-foot peak, then plunged into the valley on the other side. Another tornado, in Alabama, survived a leap off a 200-foot cliff. A North Carolina tornado kicked off its path at a 3,300-foot ridge and steamrolled down a 1,200-foot drop into Stecoah. A deadly Brandenburg, Kentucky, tornado crossed the Ohio River, and another 3 combined for a nearly continuous 121-mile track, crossing the Tippecanoe River in Indiana, where 5 people were killed in a minibus thrown from a bridge.

Witnesses provided other valuable insights. A video showed a 140-mph updraft eat away part of one tornado. The many photos of the Xenia tornado indicate its funnel cloud may never have touched the ground during its most damaging phase. At its inception, two vortices about 30 feet across wrapped around each other in crisscrossing paths. Minutes later a puff of dust scoured the ground with clear air between it and the cloud. Many witnesses swore they saw as many as six or eight vortices within the tornado. A teenager filmed the storm as it bore down on the Ohio town: The picture revealed 219-mph winds in the dust. The Xenia tornado was one of six that day that Fujita rated a rare F5 on his damage scale. That's more F5s in one day than appeared in all of the 1980s.

One survivor in Charlottesville, Indiana, looked on the bright side while picking through the rubble that was once his house. "Well, we won't have to worry about looters tonight," he said. "There's nothing left to take."

THUNDERSTORMS / TORNADO SCIENCE

A Kansas tornado in 1879 tore an iron bridge off its piers and threw it to the ground, twisting it beyond recognition. The water below was sucked from the riverbed, exposing the mud beneath. In 1990 another Kansas tornado ripped 88 train cars from a track, piling them in heaps four-deep. It was by no means the strongest tornado in the state that year.

Tornadoes are nature's ultimate wind, wrapping air and water and debris at incredible speeds around a tall, tube-like low-pressure core. This prodigious force cannot strike anywhere or at any time, but instead tends to form only under the updrafts of the most intense thunderstorms. Meteorologists brave enough to launch balloon-borne instrument packets into severe storms have clocked air rising at more than a hundred miles an hour. A rapid updraft is crucial to tornado formation, in part because it sustains the core of low pressure. The conditions that produce these monster storms were fully evident in the Jumbo Outbreak in 1974: low-level moisture and heat, jet-stream energy, and disturbances such as strong fronts.

In the 1940s, meteorologists at Tinker Air Force Base near Oklahoma City began studying tornado outbreaks, looking for forecasting clues about severe storms. Then, in 1948, a tornado struck Tinker. Maj. Ernest Fawbush and Capt. Robert Miller began applying their research, and five days later, they made a successful forecast. Their methods for finding regions likely to develop dangerous updrafts became the basis for severe thunderstorm forecasts—and tornado, large hail, and frequent lightning outlooks—issued by the National Weather Service.

Forecasters anticipate severe thunderstorms by checking for instability—the potential for warm, moist surface air to rise into a cooler environment. Often this instability is greatest in the southern reaches of a storm system, but this doesn't mean tornadoes are more likely farther south—other factors are critical as well.

Ironically, one of these factors is a layer of stability, or cap, in which temperatures increase with height. This is an inversion, a reversal of the usual temperature profile of the atmosphere, often at several thousand feet up. The cap holds convection in check long enough to allow surface air to build up moisture and heat, and hence its potential for buoyancy. The moisture must be concentrated, and on the Great Plains this is how it is often supplied: A wet tongue of air from the Gulf of Mexico forms a distinct boundary, or "dry line," next to drier air from the Southwest. Even without temperature contrast, the dry line acts like a cold front because of the difference in density between lighter moist air and denser dry air (water vapor has about two-thirds the molecular weight of other molecules in the air).

Something must break the cap to trigger convective energy. Large low-pressure systems provide the necessary disturbance. Their fronts push the surface layer out from under the edge of the cap, or they can lift both the surface and cap layers until the cap is no longer stable—no longer warmer than the layer below. Sometimes the dry line is enough of a disturbance, forming squall lines within the warm sector of a larger, low-pressure cyclone. A squall line, stretching at times hundreds of miles, creates several favored places for tornadic storms: the intersection of

From two miles out, University of Oklahoma researchers train their portable radar on a tornado near Hodges, Texas. Advanced equipment that can sample thunderstorm characteristics in and around funnels helps meteorologists determine why fewer than half of severe thunderstorms spawn tornadoes. The relative rarity of tornadoes means that even expert chasers might close in on a twister in only 10 percent of their expeditions.

the seasonal cycle of strong fronts. Large temperature contrasts make March and April peak months for the strongest tornadoes on the Great Plains, and the resurgence of fronts in October and November encourages a revival of tornadoes in autumn. In all, three-quarters of U.S. tornadoes occur from March to July.

But even a strong cold front is unlikely to trigger a tornado without the guidance and energy of a strong jet stream. A well-placed jet strengthens a low-pressure center and drives the cold front forward, which in turn plows through the warm, capped air, shocking it into convective action. Because such a jet often sweeps northeastward, the storms move northeastward as well. About two-thirds of Iowa tornadoes move in this direction, while less than 2 percent come from the east.

Crossed by the jet stream and powerful spring storm systems, the Great Plains satisfy the conditions for severe storms famously well—the region is the world's tornado alley. Without an east-west mountain chain (such as

the squall line and a front, or dry line; the center and front of the typically bow-shaped line; and the southern flank of a squall line (no competition for moist Gulf air).

Because of this need for large-scale disturbance, tornado activity waxes and wanes with the Alps in Europe), warm, moist air from the south spreads far inland in North America, meeting the cool, drier air from Canada. A century ago, surveys of U.S. tornadoes tallied about 100 a year. Now, with a denser population and better-trained observers, that count often exceeds 1,000. Oklahoma has more tornadoes per square mile than any other state—about 9 annually for every 100-by-100-mile plot.

Tornadoes also strike China, Japan, New Zealand, Australia, South Africa, Europe, and South America. England has more tornadoes per square mile than any other country, but nowhere are they as tragic as in Bangladesh, one of the few countries where they can approach the power of a Kansas whirlwind. More than 1,100 people were killed and 100,000 left homeless in a twister near Dhaka in 1989, and at least 400 were killed and more than 50,000 injured in a similar disaster in May 1996.

On May 3, 1943, Roy Hall of McKinney, Texas, watched from his porch as the western sky turned inky black. Between peals of thunder, he could hear a roar like "a distant freight train." He and his family ducked inside before tennis-ball-size hail began pelting their house. Behind the curtain of hail lurked a tornado.

The wind ahead of the twister shook the house, knocking over the west wall; then the house disintegrated. Hall was knocked on his back, and found himself looking up into the vortex. "The interior of the funnel was hollow: the rim itself…not over 10 feet in thickness and …perfectly opaque. Its inside was so slick and even that it resembled the interior of a glazed standpipe. Down at the bottom…the funnel was about 150 yards across. Higher up it was larger,

More than 1,000 tornadoes strike the United States each year, but only about one percent pack the 200-mile-an-hour winds that ravaged Albion, Pennsylvania, on May 31, 1985 (opposite). Barrie, Ontario (left), also took a direct hit from one of the 43 tornadoes touching down that day in the U.S. and Canada.

and seemed to be partly filled with a bright cloud, which shimmered like a fluorescent light. This brilliant cloud was in the middle of the funnel, not touching the sides." The twister roared past, and the family survived.

Until the late 1940s, eyewitness accounts were among the few pieces of evidence scientists had about tornado formation and structure. Eyewitnesses, however, were only able to describe the rudiments of tornado structure, such as the vigorous rotation. Better observations were necessary to understand the connection between the tornado and its parent storm. But without better forecasts, more thorough thunderstorm knowledge, and more powerful instruments, meteorologists could not get safely close enough to make good observations.

With the advent of Fawbush and Miller's new guidelines and new instruments, meteorologists began to probe thunderstorms in earnest. Radar and observation networks revealed a thunderstorm-size atmospheric rotation, called a mesocyclone, in the vicinity of tornadoes. Then came a breakthrough when a tornado swept into Fargo, North Dakota, in 1957. Drivers on the highway leading into the city passed a dark, spinning cloud moving at about 20 miles an hour and warned local authorities of a tornado. The early alert gave residents time to seek shelter, but some decided instead to bring out their cameras to record the event. The

disaster killed ten people and ruined more than a thousand houses, but it remains one of the best-documented tornadoes in history. The photos showed that the initial cause for the warning was not a tornado but a lowered, rotating base of the cloud's updraft region. The entire cloud tower was a giant parent vortex. The precipitation-free protrusion, now called a wall cloud, is a precursor to tornadoes.

Visual proof and radar evidence helped meteorologists identify the ideal thunderstorm for producing tornadoes—the supercell. Many tornadoes come from multicellular storms with two or three updraft regions. The supercell, on the other hand, can have a single rotating column of rising air. It is perfectly adapted to perpetuating a powerful updraft, and hence a tornado, by blocking winds aloft and by dropping rain shafts outside the mother vortex. Such storms can last for several hours and produce multiple tornadoes.

The supercell rotation, or mesocyclone, begins around 14,000 feet or more in altitude and eventually extends through a significant depth of the storm. The twist inherent in strongly veering winds that increase with height causes the rotation by rolling air into spirals, which are pulled upright by the storm. Some storms have rotation pairs—a counterclockwise cell (as viewed from above) and a clockwise cell. These twins can split: The counterclockwise cell, which is often dominant, moves to the right of prevailing winds; the other moves slightly left of prevailing winds. The former cell is a prime tornado threat; the latter often brings hail.

By the early 1970s, improved roads and forecasting techniques enabled scientists to hunt for thunderstorms and try to document on film the evolution of a tornado from the parent storm's rotation. Then they began to measure the thunderstorms with instruments placed in front of storms (or launched on balloons and rockets). New supercomputers emerged, capable of using the measurements to model thunderstorms. The combination of observations and models has helped meteorologists inch to the brink of explaining how a tornado forms.

Doppler radar, lately in portable form, is one of the principal instruments in this quest, because it can measure movement in the atmosphere. With Doppler, scientists can clock high-speed rotations in the middle and upper regions of supercells. Chasers in the field, complementing the radar readings, observed important features of supercells. For instance, they found that tornadoes tend to form on the southwest edge of a storm, in the wall cloud and adjacent to the rain-cooled downdraft wrapping around the updraft. They often saw no lightning or rain near the tornado, debunking theories that these phenomena spawn tornadoes. And the clockwise tornadoes they found on the Plains helped prove that the spin of the tornado or the parent storm was not primarily due to the earth's rotation. This is true even though all but a few Northern Hemisphere tornadoes are counterclockwise and Southern Hemisphere tornadoes are clockwise. Chasers

and radar operators found that supercells often telegraph their tornadic intentions about 20 minutes ahead by forming a small vortex inside and along the edge of the rotating updraft. Called a tornado vortex signature, this tube can spin at 150 miles an hour and stretches high enough to be spotted by Weather Service Doppler radar (low-level features are often blocked by the horizon).

Scientists interested in the life cycle of tornadoes first tried their chasing skills on the more benign waterspouts, and actually flew airplanes through some. These water-borne whirlwinds average winds of only 50 miles an hour, and can form in fair weather. Many weak tornadoes are waterspouts that move inland. They have a similar life cycle to tornadoes, which often peak as wedge-shaped funnels and then become erratic and ropy near the end.

In waterspouts, the wind shear below a cloud's updraft creates little vortices that spin up and stretch into whirlwinds. By the 1980s, chasers in the High Plains found similar land-based tornadoes spinning up in the wind shift along gust fronts, then stretching vertically under the updraft of a storm. After years of chasing supercells, the discovery of weak tornadoes under rapidly growing cumulus—clouds with no rotation—came as a surprise. Howard Bluestein, a scientist at the University of Oklahoma, dubbed the nonsupercellular tornadoes "landspouts."

Tornado formation without supercells suggested that the low-level circulations below a mesocyclone play an important part in triggering a supercellular tornado. Computer models showed the significance of the rain-cooled

Walls shattered and the roof collapsed when a twister ripped a house in Lancaster, Texas, but pictures clung to a wall and a line of trophies held their ground. The aftermath of the 1994 storm testifies to the capricious powers of tornadoes. Anchoring construction to resist uplift and giving more aerodynamic lines to roofs can help many houses withstand all but the most violent tornadic winds.

downdraft hitting the ground near the wall cloud. Other computer simulations suggested tornadoes are the result of strong, low-level rotations that reveal themselves by sucking down air from within the mesocyclone. But other theorists held that the high-level rotation of the tornado vortex signature eventually lowers to the ground, stretching and accelerating the way figure skaters speed their spins by drawing in their arms.

Figuring out what happens at low levels to initiate the formation of a tornado was one of the goals of Project Vortex, a series of storm chases in 1994 and 1995. Scientists from the University of Oklahoma, the National Severe Storms Laboratory, and other groups teamed up to surround thunderstorms with a horde of instrumented vehicles. They brought along the most advanced portable Doppler radars as well—including aircraft-borne and truck-mounted dishes. Scientists watched from a radar airplane only a thousand feet in the air as tornadoes ripped up asphalt and hail pelted their colleagues' vans.

The new fine-scale radars revealed spiraling bands within one cloud, and a tornado vortex tube that stretched from the ground to near the top of another storm. The resolution on some imagery was good enough to show the downdrafts that supercomputer modelers had suggested would be caused by low-level rotation, implying that tornadoes "spin up" rather than "touch down." Inside one funnel researchers could see concentric rings of debris that may have been centrifuged outward by rotation. Finally, researchers were actually looking inside a tornado, recording what once was the sole province of a few survivors like Roy Hall.

THUNDERSTORMS
TRI-STATE TORNADO OF 1925

One witness mistook it for a rolling fog at the bottom of the thunderstorm. Many in the railroad towns thought the roar came from a steam engine belching black smoke. When the whirlwind drew closer, people saw the strange yellows, greens, and reds in the clouds, then heavy hail, then the chairs, chickens, boards, books, and shingles swirling in the air. Then they all knew what the heavy humid air had made old-timers suspect all day.

Cloaked in a storm cloud that brushed the ground, the tornado took most of its victims by terrible surprise. Weather historian David M. Ludlum calls the Great Tri-State Tornado of March 18, 1925, the "greatest physical force of a tornadic nature ever to develop within the United States." Certainly, nothing like it has happened since the murderous twister ended its record three-and-a-half-hour spree, killing 689 people and injuring more than 2,000. Four towns were nearly completely destroyed.

This longest-lasting, deadliest tornado in American history was also one of the best traveled, plowing a continuous swath of damage from a farm near Redford, Missouri, to several miles east of Princeton, Indiana—an amazing 219 miles. On average the whirlwind was a quarter-mile wide, and at one point it raced at 73 miles an hour.

The frontal storm that spawned the tornado made a long journey of its own from Alberta, Canada, on March 16, to Texas and then into southwestern Missouri by the morning of March 18. A cold front extended south into Iowa and Kansas. That day, moving east-northeast at about 30 miles an hour, the storm spawned 8 tornadoes, killing 97 people in Kentucky, Tennessee, and Alabama.

The tornado sped 70 miles an hour at its start and leveled Annapolis, Missouri. But the tornado caused its worst damage after it crossed the Mississippi River, when it caught up with the low-pressure nexus of the frontal storm. The twister slowed to less than 60 miles an hour and in 40 minutes killed 541 people as it smashed town after town along a railroad. At 2:26 the tornado virtually leveled Gorham, Illinois, killing dozens of people. At the schoolhouse, children saw the skies turn "black, like night." Then the tornado hit the building—the walls collapsed and the floors gave way. Most of downtown Gorham was demolished—a cow landed in the town's restaurant. The tornado then sped into Murphysboro, destroying nearly 100 blocks before fire consumed another 70. Many people who made it to their basements were killed when their houses collapsed. People trapped under the rubble of a hotel were consumed by fire. It was impossible for fire companies to negotiate the streets littered with wreckage, and many hydrants were broken. Minutes later, the tornado destroyed the town of Parrish. The houses that weren't reduced to rubble were wrenched from their foundations.

Students in Crossville, Illinois, watched from their school windows as the tornado split in two, then merged again in their neighborhood. They saw houses vault into the whirling black cloud or simply collapse. Across the Wabash River in Indiana, Griffin was a complete loss—not a building withstood the onslaught. Survivors and victims alike were covered with mud from the riverbanks. One town wag saw

In an era without advance storm warnings, the Great Tri-State disaster of March 18, 1925, produced the worst tornado tragedy in American history. That day, one tornado crossed 3 state lines and killed nearly 700 people. The desolate aftermath of the warm spring day in Griffin, Indiana, where more than two dozen people died, reminded local veterans of the battle-scarred landscapes of Europe after World War I.

the tornado and ran for the train station: "As I took hold of the doorknob that storm just naturally jerked the station right out of my hand."

Shortly before lifting into the sky, the vortex took four miners by surprise as they motored home near Princeton, Indiana. They were thrown to the ground as their car flew out from under them. The bundle of paychecks in the office of the Heinz Ketchup plant nearby, on the other hand, was thrown more than 50 miles.

The aftermath was grisly, with bodies flung into trees, chickens plucked clean, and cows impaled by two-by-fours. In Gorham, five hours passed before surgeons reached town to aid the wounded. The railroad organized a train to rush victims the 50 miles to Cairo, Illinois, the nearest hospital not already filled with casualties.

Many of the region's men were coal miners, and they emerged from the earth stunned by the wreckage. Veterans of World War I regarded the colorless landscape of defoliated trees and bare stumps that tragic afternoon with a sad sense of recognition. Only this time, the dead were mostly women and children.

Severe wind damages covered an area larger than Hurricane Andrew's swath in Florida. Even today, with advanced warning systems, the speed and humble appearance of the Tri-State Tornado would make it a formidable killer. As snow flurries began to fall in the area, a new era began for the residents. In Murphysboro, where the major industries were destroyed, one resident later said, "When we talked, we talked 'before the tornado' and 'after the tornado.'"

HURRICANES

In 1274, Kublai Khan sent an invasion force of 40,000 men to conquer Japan. But a November storm, probably a typhoon, drowned 13,000 of the invaders, and what was left of the Mongol fleet hobbled home.

A divinity among men, the khan was undeterred. At the height of the monsoon season of 1281 he sent 140,000 warriors to fight again. On August 15 his 4,400 ships massed at Hakata Bay. But again the winds sided with the samurai: Another storm scattered the invaders, its massive tide dashing ships against rocks. Chroniclers say the wreckage was so thick with bodies and boards that a person could walk across the bay.

Storms like the kamikaze—the divine wind that delivered Japan from Kublai Khan—have overwhelmed humanity for centuries. Not by accident did Spaniards adopt an indigenous word used for evil spirits and weather gods, *huracan,* to name the tropical scourge that sent thousands of their ships to the bottom of the Caribbean. Today these storms go by various names: hurricanes, typhoons, and more. The name does not matter. Wrote one early settler on Martinique: "They are the most horrible and violent tempests one can name; true pictures of the final fire and destruction."

If the thunderstorm is an airborne mountain, then the hurricane is an oceanic range. Its characteristic coil, now familiar from satellite and radar imaging, was early recognized as a snake of the skies, earning it the name "cyclone."

MIKE LACA/WEATHERSTOCK

Every second, a hurricane generates the energy released by ten Hiroshima-size atom bombs. Thus, with a Caribbean war looming in 1898, the head of the U.S. Weather Bureau went to the Commander-in-Chief to warn the Navy of the hurricane threat. President McKinley was impressed. "I am more afraid of a West Indies hurricane than of the entire Spanish fleet," he declared, establishing a pioneering storm-warning network.

Perhaps McKinley remembered the Navy's attempt to thwart German imperialism in Samoa. In 1889 six ships of the two nations squared off in the harbor of the island, when a typhoon blew in. Every ship sank—though thanks to a valiant rescue effort by the Samoans, only 150 sailors drowned. The parties promptly retreated to negotiations, and war was averted.

More than half a century later a crafty Adm. William F. "Bull" Halsey maneuvered an

Hurricane Kate's helter-skelter winds, which arose from the large-scale circulation around an eye of calm, lash the coastline of Key West, Florida, in 1985 (opposite). Seen from space, such an eye forms the open bowl at the center of Hurricane Bonnie in 1992 (left). The size of a hurricane's eye, whether five or fifty miles across, says little about the storm's relative strength, but as the trademark of a mature tropical cyclone, the eye symbolizes nature's awesome ability to sculpture order out of chaos.

American fleet around a typhoon that battered the Japanese. The helpful storm was dubbed Task Force 00. His luck ran out in December 1944 when he sailed his Third Fleet into the path of another typhoon. Three destroyers capsized in 80-foot waves, killing 790 American sailors. The admiral escaped with a rebuke from his superiors, only to repeat the mistake in 1945. Perhaps the Japanese could have told him a thing or two about typhoons, had they not kept secret the fact that their own admirals had suffered comparable losses in a 1935 typhoon.

Whereas Halsey's audacity failed him, the instrument-flying abilities of Col. Joseph Duckworth advanced hurricane understanding. A veteran pilot instructor, he heard a hurricane was approaching Galveston in 1943, grabbed a navigator, and took off in clear skies. The two-seat trainer was "tossed about like a stick in a dog's mouth." The altimeter proved useless in the extreme pressure drop, and the radio was filled with static. Still, Duckworth became the first to peer into a hurricane's eye.

Routine flights followed. Air blasts sheared rivets off wings, and water sprayed through fuselage seams. One hurricane hunter reported, "One minute this plane, seemingly under control, would suddenly wrench itself free, throw itself into a vertical bank and head straight for the steaming white sea below. An instant later it was on the other wing, this time climbing with its nose down at an ungodly speed.… I stood on my hands as much as on my feet."

On a flight through Hurricane Edna in 1954, while wondering at the great amphitheater of clouds below, the legendary CBS reporter Edward R. Murrow gained the perspective that should have been Kublai Khan's: "If an adequate definition of humility is ever written, it's likely to be done in the eye of a hurricane."

HURRICANES / GALVESTON 1900

"With a raging sea rolling around them, with a wind so terrific that none could hope to escape its fury, with roofs being torn away and buildings crumbling…men, women and children…huddled like rats in the structures. As buildings crumpled and crashed, hundreds were buried under the debris, while thousands were thrown into the waters, some to meet instant death, others to struggle for a time in vain, and yet other thousands to escape death in miraculous and marvelous ways." This is an eyewitness account of the Galveston hurricane of 1900, the deadliest natural disaster in U.S. history. It killed more Americans than the Chicago fire, San Francisco earthquake, and Johnstown flood combined, yet today is all but forgotten.

The storm killed at least 8,000 people. It cut Galveston Island off from the Texas mainland and submerged it under the sea. In the city of Galveston, on the eastern end of the island, the storm killed some 6,000 men, women, and children and left 5,000 more battered and bruised. It swept away 2,636 homes and reduced to rubble at least 1,000 more. Not one building escaped damage; every survivor had faced death.

Galveston had been "a city of splendid homes and broad clean streets," wrote Clarence Ousley of the *Tribune*, "a city of oleanders and roses and palms; a city of the finest churches, school buildings, and benevolent institutions in the South; a thriving port…; a seaside resort…; a city of great wealth and large charity." After the storm, it was "a city of wrecked homes and streets choked with debris sandwiched with six thousand corpses; a city…with the slime of the ocean on every spot and in every house; a city with only three churches standing, not a school building or benevolent institution habitable; a port with shipping stranded and wrecked many miles from the moorings…a rolling surf three hundred feet inward from the former water line…lapping an area of total destruction four blocks deep and three miles long…a city whose very cemetaries [sic] had been emptied of their dead."

AMERICAN RED CROSS, COURTESY PATRICK E. HUGHES

The storm that destroyed so much of Galveston was born west of the Cape Verde Islands about August 27. By September 4, it was moving northward over Cuba. It became a full-blown hurricane as it raked the Florida Keys on September 5. On Friday the 7th, the Weather Bureau in Washington notified Isaac Cline, chief of the Galveston office, that storm warnings had been extended to Galveston.

That afternoon, Capt. J. W. Simmons of the *Pensacola*, en route from Galveston to his home port, watched the barometer fall to 28.55 inches as he struggled to stay afloat in towering seas. He had never seen the glass that low before.

At daylight Saturday morning, early rising Galvestonians went down to the beach to watch the thundering surf crash farther and farther inland. Already, breakers had smashed beach-side buildings and destroyed a streetcar trestle.

A great hurricane destroys Galveston, Texas (opposite), in September 1900, throwing thousands of residents into raging, storm-tossed waters. Refugees seeking sanctuary from the storm quickly filled Sacred Heart Catholic Church (left), but waterborne wreckage from hundreds of homes soon battered down the structure's south and east walls; no one knows how many people died there.

At 10:10 a.m., Washington warned Cline the storm center was expected to make landfall west of Galveston, putting the city in the hurricane's right semicircle. There the winds would be strongest, exposing residents to the threat of a deadly storm surge.

By noon, the wind was blowing out of the northeast at more than 30 miles an hour and increasing steadily. The barometer was dropping rapidly and heavy rain was falling. The Gulf of Mexico was already three to five feet deep in streets near the beach, and cottages in the eastern end of the city were breaking up. Meanwhile, the rising water of Galveston Bay had submerged the wagon bridge and train trestles connecting Galveston to the mainland and were flooding inland over the wharves on the north side of the city. It was too late to get

off the island. "Gulf rising rapidly; half the city now under water," Cline wired Washington at 2:30, adding the city was going under fast, and "great loss of life must result."

Between 4 and 5 p.m., the waters of the bay met those of the gulf, submerging the city. By 5:15, the wind was gusting to 100 miles an hour, rolling up tin roofs and toppling telephone poles. Huge chunks of masonry were crashing into flooded streets in the business district, while near the beach, homes were being pounded by wind, water, and wave-tossed wreckage. Flying debris from disintegrating buildings was killing people in the streets, while whirling wreckage pinned others underwater.

Isaac Cline was standing in 8 inches of water at his elevated front door about 6:30 when the water suddenly rose above his waist. Gulf water was now 10 feet deep in the street and 15.2 feet above sea level. The water rose another 5 feet in the next hour.

By 7 p.m. the sky was dark and the wind was howling at an estimated 120 miles an hour. Waterborne wreckage and a railroad trestle a quarter mile long were battering Cline's house. At 7:30 the wind shifted to the east, blowing with even greater fury, and the house fell. It carried with it about 50 people who had sought shelter there. As the house capsized, Joseph

Cline grabbed his brother's older girls and jumped through a second-story window. Isaac Cline, his wife, Cora May, and his daughter Esther were carried under when the house rolled over. Cline lost consciousness. When he awoke, he was above water. Esther was nearby, but Cora May was gone.

A few minutes later, Cline saw his brother and his other two children atop the floating house and joined them. For the next three hours they scrambled from one piece of wreckage to another as the house broke up. The rain fell in torrents, while the wind hurled timbers and other missiles past their heads. The brothers huddled with their backs to the wind and the children in front of them, holding planks behind them to ward off the deadly flying debris.

The wreckage they were clinging to grounded about 10:30 p.m. The exhausted survivors climbed over debris mounds to a still-standing,

two-story house and crawled in an upstairs window. Almost a month later, the body of Cora May Cline was found under the wreckage that had carried her family to safety.

When her home went down, Mrs. William Henry Heideman, who was pregnant, was swept away on a roof from a wrecked cottage. The roof struck other debris and she was hurled off—only to land in a steamer trunk bobbing by on the surging water. The trunk hurtled along until it smashed against the wall of the Ursuline Convent, where it was pulled inside. Mrs. Heideman's son was born a few hours later.

The rambling wooden buildings of St. Mary's Orphanage stood on the beach 3 miles west of the city. They housed 10 Roman Catholic nuns, 93 orphans, and a workman. The water was already 3 feet deep by 10 o'clock Saturday morning. Later, as the storm worsened, the sisters took the children to the chapel

AMERICAN RED CROSS, COURTESY PATRICK E. HUGHES

on the first floor of the girls' building. They prayed there until the rapidly rising water drove them to the second floor. From there they watched the boys' building break up. Sometime between 7 and 9 p.m., the roof of their building collapsed. With death threatening, the sisters tied the children to their waists.

After the storm, only a few bricks marked the site of the orphanage. Ninety bodies were found nearby. One nun still had nine small children tied to her body; another was holding a child in each arm. Only three orphans, all boys, survived. Somehow they got out of the building and onto a tree sweeping by on the water.

The dawn brought new horror. All that remained of thousands of homes and the people who had lived in them was a great mound of wreckage jammed with bodies the storm had piled up around the business district. Outside this barrier, everything had been swept away.

Many bodies and many of the living were naked or near-naked, their clothes ripped off by nails, jagged pieces of wood, broken glass, and flying debris. Hundreds of whole families had perished. As many as 10,000 survivors were homeless. Galveston Bay was clogged with human bodies, and the corpses of cows, horses, chickens, and dogs were everywhere. By Monday, makeshift morgues were jammed with corpses. The weather was hot, and the health of the living demanded that quick action be taken. So on Tuesday, Galveston became a city of funeral pyres. Dark smoke hung over the island for weeks as the grim task continued.

Ironically, the hurricane itself saved many survivors—and showed them how to prevent a similar disaster in the future. The barricade of wreckage thrown up by the storm had acted as a breakwater or seawall, shielding the buildings and people behind it from the destructive force of racing floodwaters, crashing waves, and whirling wreckage. So Galveston built a seawall.

On the night of August 16, 1915, the city was hit by a similar hurricane that closely followed the track of the 1900 storm. This time, only eight people died in Galveston. The city had learned its lesson well.

Today, millions of Americans live in low-lying coastal communities from Texas to Maine; in summer, the number of people who live and play in areas that could be swept by winds and waves is much greater. Unless they are willing and have time to evacuate before a major hurricane strikes, not even weather-satellite warnings can prevent a disaster that could dwarf the horror of the Galveston hurricane.

Patrick Hughes

HURRICANES / ANDREW

Soon after a New England hurricane in 1821, William Redfield decided to visit his relatives. A frugal man of limited means, he walked clear across Connecticut. Redfield noted how broken branches and uprooted trees lay in one direction near home, yet farther west lay in the opposite direction. He realized that the storm had been a giant whirlwind, and his observations revolutionized our understanding of tropical cyclones.

The spiral of a hurricane is a paradigm of tightly woven order—a restructuring of nearly a million cubic miles of chaotic atmosphere. Immense power is needed to do this, but a hurricane forms in a region devoid of the frontal boundaries and jet-stream energy that build a midlatitude storm. Lacking these helpful contrasts, a hurricane makes its own violence from vast expanses of warm seawater and humid air.

A hurricane gets its power by forming a massive heat engine to sort out the differences between the warm sea surface and the cool atmosphere high above it. The fuel line for the engine is a cyclonically rotating inflow, hundreds of miles wide, that gathers humid air in a surface layer about 10,000 feet thick. The inflow creates a banded coil of storms and accelerates as it nears a low-pressure center. Here, before it can reach the eye of calm at the center, the humid air is forced into a ring of relentless updrafts lifting a billion tons of air at a thousand feet per minute. Vapor condenses in the updrafts, releasing heat high in the atmosphere.

As a heat engine, the hurricane is extravagant. In two days it can release enough energy to generate all of the electricity the United States uses in a whole year. Yet the storm uses less than one-tenth of that heat release to accelerate winds near its center to 150 miles an hour and drive waves 20 feet high in a storm surge.

Despite its excessive strength, the hurricane is a temperamental engine that reluctantly forms and quickly conks out in adverse conditions. Thus it was the exception, not the rule, when a common atmospheric disturbance over the tropical North Atlantic developed into a nightmare called Andrew.

On August 14, 1992, a patch of clouds moved off the west coast of Africa and into the Atlantic Ocean. The disturbance was typical of the region: Every three or four days in the Atlantic's June-November hurricane season, a wave appears in the easterly trade winds blowing from Africa toward the Americas. This easterly wave encourages low-level air to converge and rise east of its crest. Only about one in seven of these disturbances becomes a hurricane; similar easterly waves help seed storms in the Pacific.

Like all hurricane seedlings, this easterly wave would grow only in perfect conditions. First, the sea surface had to be 78°F or warmer over a wide area and overlaid with warm, moist air. The warm water had to be at least about 200 feet deep, because a hurricane will cool the immediate surface by several degrees as it whips up waves and taps heat. Some hurricanes have actually died in the cooled wakes of other storms. Normally, however, easterly waves find the ocean plenty warm when they pass the Cape Verde Islands. Sure enough, in this region, the first successful wave in 1992 began forming a rotating inflow around a central low-pressure area. Officially, the nascent

Hurricane Andrew crosses South Florida on August 24, 1992, and enters the Gulf of Mexico. Only slightly diminished, the storm regained strength over the warm waters and headed for Louisiana, where it made a final assault on land. The hurricane had previously packed 140-mile-an-hour sustained winds in Dade County, making it the third-strongest landfall in U.S. history.

"Cape-Verde type" hurricane was now a tropical depression, less than 48 hours out of Africa.

By now a chain reaction was in full swing. The spiraling inflow of warm humid air, drawn in by low pressure at the center of the depression, met near the center and ascended. Aloft, much of the vapor condensed, releasing heat that expanded the updraft. The buoyant air then rose even faster until it hit the tropopause, an atmospheric ceiling of stable air at about 40,000 feet. Here the air formed a canopy of icy cirrus that spilled downdrafts outside the storm. The icy exhaust at the top pumped air away, lowering the pressure at the center, which in turn drew in warm humid air more vigorously.

To sustain this chain reaction in August 1992, winds above the ocean met another condition for hurricanes: They were relatively uniform throughout the troposphere, from the sea surface to about 50,000 feet. If upper-level winds are too fast or blow in a different direction from the low-level easterlies, the seedling's

updraft is dispersed prematurely. Without updraft heating near the core, the critical chain reaction grinds to a halt. Few hurricanes form in November over the Atlantic because the chain reaction is disrupted by the migrating jet stream, which rips the tops off tropical storms.

The seedling benefited from all the right conditions—at first. As the pressure continued to drop, the air accelerated faster inward. Winds topped 37 miles an hour, meaning the storm earned tropical storm status and a name. Tropical Storm Andrew debuted on August 17.

Andrew had nearly a two-thirds chance of making hurricane status, defined in the Atlantic by 74-mile-an-hour winds. A mature hurricane sustains such winds out to an average radius of 70 miles from the center, only half the total width of the storm's circulation. Some spread gale-force (39-mph) winds more than 600 miles in diameter, but size is not the best indicator of strength. Indeed, like Camille in 1969, and Cyclone Tracy in 1974 (one of Australia's worst disasters), Andrew was relatively small, with gale-force winds no more than a hundred miles from the eye.

Thanks to a big break in the winds around it, Andrew developed unimpeded. On the 20th, it dodged a low-pressure area. The low began spewing wind shear at Andrew but went north before causing a weakening. Its southern edge may have actually enhanced the storm's upper-level exhaust system. Shortly after sunrise on the 22nd, Andrew was a hurricane.

At this point, a long bank of high pressure lay to the north: Andrew began a nearly due westward march to Miami. Had the high pressure not extended off the coast of the southeast United States, Andrew might have recurved north and then east over the North Atlantic. It would have died over cold water. Unfortunately for the Bahamas, Florida, and Louisiana, this inevitable poleward recurvature came too late.

As a hurricane, Andrew not only took a different tack, it took on a different look. Its heat engine was more powerful, and accelerating winds reached an energy limit that forbade them to approach the center. This resulted in the mark of a mature hurricane, a clearly developed eye. This calm amid the tempest can be as few as five or as many as fifty miles wide, and is often filled with seabirds trapped by the storm.

Air sinks within the eye of the storm, sometimes clearing the sky, and this clear calm frequently fools people into believing the storm has completely passed. During the Great Miami Hurricane of 1926, people emerged from hotels and apartments during the eyewall passage. Some headed to the beach for a swim; others surveyed the destruction. Forty minutes later,

the renewed fury of the storm drowned more than a hundred of them.

Mariners are less likely to be confused by blue sky in a hurricane, in part because the waves inside the eye can be chaotic. The skipper of a ship caught in the eye of a China Sea typhoon in 1869 wrote that with "one wild, unearthly, soul-chilling shriek the wind suddenly dropped to a calm." The waters "piled up on every side in rough, pyramidal masses, mountain high [and] boil and tumble as though they were being stirred in some mighty cauldron."

In the wall around the eye, only a few thousand feet above the ocean's surface, winds of the storm are at their peak. This is one of the many differences between warm-core hurricanes and cold-core midlatitude storms. The latter get progressively stronger with height because their pressure contrasts increase with height. Sustained winds at the bottom of the hurricane, on the other hand, are usually close to two-thirds as strong as the peak winds, and a little stronger still when the hurricane is over the relatively friction-free ocean. Gusts may carry peak velocities completely to the surface.

Andrew was able to produce sustained surface winds of 155 miles an hour in its eyewall. Part of the secret of Andrew's success was the speed with which its chain reaction worked. Rapid deepening of the central eye pressure seems to go hand in hand with strong hurricanes. In Pacific Typhoon Forrest, one of history's strongest storms, pressure plunged from 28.82 to 25.87 inches in less than a day as winds climbed from 72 to 165 miles an hour. The strongest Atlantic hurricanes did the same: Gilbert in 1989, Allen in 1980, and Camille in

After Andrew's passing, DC-3s take to the skies to spray a burgeoning mosquito population and quell threats of disease in devastated South Florida. In Coral Gables (below), a pile of powerboats accounts for only a handful of the 15,000 Dade County craft battered by the storm.

1969. Similarly, Andrew's central pressure deepened rapidly. In its first day and a half as a hurricane, Andrew passed over 86°F water. Its central surface pressure fell nearly 8 percent and its eyewall tightened to only 5 miles across, enough to double sustained winds near the center. The winds, and a lowest central pressure of 27.22 inches, made Andrew a Category-5 storm.

Two days later, Andrew's eye passed over Eleuthera Island and then the Berry Islands in the Bahamas. Here the storm whipped up a 23-foot storm surge, but it also met its first real challenge: the Great Bahamas Bank. Though the water here is warmer than in the deep ocean, it is only about 5 feet deep at low tide. Andrew's heat engine wasn't efficient enough to run on this shallow fuel tank. In just 3 hours, the eye pressure rose and the strongest winds slowed to 115 miles an hour in the eyewall.

As it weakened, Andrew suffered a common crisis: A concentric wall of updrafts formed outside the original eyewall, stealing inflow from the storm's central engine. Many hurricanes never recover from this development, and the uncertainty about Andrew's strength during this concentric eyewall replacement was an encouraging sign as Florida landfall became inevitable. But Andrew's second eyewall quickly tightened around the first. Before dawn on the 24th, the storm had a single eyewall again and 140-mph winds. The eye crossed land east of Homestead Air Force Base at 5:05 a.m. local time. It was moving close to 20 miles an hour, twice the usual speed of hurricanes.

Such speed was in some ways fortuitous. Hurricanes move with the total current of air around them, and fast currents prevent storms from building up the waves they plow ahead of them. Lingering storms, like the giant Felix in 1995, can erode shorelines with relentless waves. Slow storms can also cause tragic mudslides and floods.

Andrew, small and fast, dropped only about 7 inches of rain in Florida. Its localized storm surge reached a record 16.9 feet along the Bay of Biscayne but missed such vulnerable communities as Miami Beach. Thanks to timely warning, more than a million people had been evacuated, and only 43 people died.

But Andrew's speed also meant that its winds compromised little when moving ashore, and the westward winds in the north eyewall were intensified by the storm's forward motion. Fast storms penetrate farther inland than slow storms. Speedy Opal brought gusts as far north as the Atlanta suburbs in 1995. Powerful Hazel,

in 1954, kept up a 60-mph clip after charging into North Carolina, and its eye remained intact as far north as Toronto.

Land stops hurricanes even more efficiently than cold water. A storm like the March 1993 blizzard can intensify rapidly over land. But the atmosphere over land provides little vapor to fuel the greedy heat engine of a tropical storm. The updraft machinery conks out and the storm glides in for a crash with midlatitude air: The eye fills in and the organized inflow disintegrates. The storm might be swallowed by a midlatitude storm, or hurricane circulation might be strong enough to suck in a cold sector of its own, creating a frontal boundary and transforming into a midlatitude storm.

Hurricane Andrew would meet this fate, but not by speeding across swampy southern Florida in 4 hours. Weakened but undeterred, Andrew regained 120-mph winds over the Gulf. Waves reached 30 feet in Terrebonne Bay,

Before and after Andrew's landfall (opposite and below): The lighthouse at Cape Florida, a survivor of hurricanes since 1846, withstood Andrew's 8.7-foot storm surge and 130-mile-an-hour winds, but some 325 acres of Australian pine surrounding the structure were less hardy. Mangroves along the shore also suffered heavy wind damage, handing non-native vegetation an opportunity to replace them.

CAMERON DAVIDSON

SAFFIR-SIMPSON SCALE

Category	Central Pressure (inches)	Storm Surge (feet)	Wind Speed (mph)
■ 1 Minimal	>28.94	4-5	74-95
■ 2 Moderate	28.47-28.94	6-8	96-110
■ 3 Extensive	27.88-28.47	9-12	111-130
■ 4 Extreme	27.17-27.88	13-18	131-155
■ 5 Catastrophic	<27.17	>18	>155

and probably topped 65 feet elsewhere as the storm finally turned northward and came ashore 20 miles west of Morgan City, Louisiana, on August 26, killing 8 people.

Andrew's journey was at an end. Less than a day after landfall, it was no stronger than a tropical depression. But like many former tropical cyclones, Andrew had a bit more to do: get rid of its copious moisture. In 1969, when remains of Hurricane Camille met a front over the Northeast, thunderstorms deluged the James River basin in Virginia. More than a hundred people died in the sudden nighttime floods and landslides in the mountains. A hurricane's intensity has little relationship to its flooding potential. Tropical Storm Alberto, far weaker than Andrew, lingered in weak steering winds over Georgia in 1994. Alberto kept pulling in moisture from the Gulf, dropping some of the worst rains in Georgia history. Flash floods broke through more than 100

dams, killing 30 people and causing more than 500 million dollars in damage.

Rain is one of the lasting legacies of a hurricane. Arid regions like the west coast of Baja California sometimes get a year's worth of rain from one hurricane. But torrential downpours can kill people en masse. In the same season that Alberto submerged towns in the river valleys of Georgia, mudslides from Tropical Storm Gordon killed a thousand people in Haiti.

Andrew dropped a respectable 10 inches of rain far inland in the Southeast, nothing like the 27 inches that fell in Americus, Georgia, two years later from Alberto. But back in Florida, more than 25 billion dollars in damage lay strewn across the landscape. Had William Redfield been able to pick through the wreckage of this costly natural disaster, he would have seen familiar patterns in the splayed trees and battered buildings. Nature's colossal spiral yet again had left its deadly mark on the land.

HURRICANES / GEOGRAPHY

arth's seven hurricane basins are prolific storm factories. Had Columbus realized his voyage to the Indies began at the height of hurricane season in the Northern Hemisphere, he might never have braved the voyage in August 1492. Given the averages, he was lucky he made a successful round-trip.

Hurricane basins are more than 4° latitude from the Equator—any closer and the effects of the earth's rotation are too weak to turn winds cyclonically. The basins reach poleward only to about 30° latitude, though storms often venture beyond this boundary before dissipating. Seasonally the basins have water about 80°F or warmer and light midlatitude westerlies.

The king of tropical-cyclone basins is the North Pacific west of the International Dateline. Here, expansive warm waters help typhoons develop very low central pressures. On October 12, 1979, Supertyphoon Tip set a record for lowest central pressure with 25.69 inches; sustained winds reached 180 miles an hour. Typhoon season peaks from June to December, but storms occur in any month. The average of 26 tropical-storm-strength cyclones, of which 16 achieve typhoon status, is more than twice the Atlantic averages of 10 tropical storms and 5 hurricanes.

The low Atlantic averages belie the productivity of easterly waves from Africa. Some of them initiate tropical storms in the eastern North Pacific, where an average of 9 hurricanes form each year. A few Atlantic hurricanes have crossed Central America to become Pacific hurricanes; one even rarer storm crossed from the Pacific and became an Atlantic hurricane.

Most eastern Pacific storms break up before reaching 145°W. A few recurve northeastward to make landfall on Mexico. In all, the North Pacific hosts more than half of the world's average annual production of 103 tropical storms and 45 hurricanes. Despite its seemingly precarious position amid hurricanes and typhoons, Hawaii rarely sees these storms. Usually a large high-pressure area and cool waters protect the state. In 1992, Hurricane Iniki became only the fifth storm since 1950 to affect the islands.

Less active basins include the northern Indian Ocean, the southwest Indian Ocean, and the two basins flanking Australia in the South Pacific. As befitting the Southern Hemisphere, Australia's cyclones peak between January and March and spin clockwise, as do the cyclones of the southwest Indian Ocean.

But averages mean little when lives are at stake. In 1992, three typhoons blew into Guam. In 1993, a record 18 tropical storms made landfall in the Philippines and 7 hit Japan in 6 weeks. In 1995, after a long stretch of quiet hurricane seasons, the Atlantic Ocean flared up with a near-record 11 hurricanes and 19 tropical storms. A record 5 named storms—Humberto, Iris, Karen, Jerry, and Luis—shared the Atlantic at the same time. Only 1969 (12 hurricanes) and 1933 (21 storms) were busier seasons.

Meteorologist William Gray, of Colorado State University, tries to pinpoint the factors that signal a strong or weak season. In 1995, nearly every factor he included in his forecast calculations encouraged hurricanes. One factor is rainfall in the western Sahel and along the Gulf of Guinea. Drought in the region coincided with quiet Atlantic seasons in the 1970s and '80s, but the rains were near normal in 1995. A second factor is the shift every 12 to 16 months in winds

Sunlit cloud tops swirl about a well-defined eye as Typhoon Odessa prowls the North Pacific Ocean in 1985. The Pacific's numerous typhoons, which can grow to several hundred miles in diameter and reach nearly 200 miles an hour, present a year-round threat to coastal regions. While relatively fewer in number, the Atlantic hurricanes born during a June-to-November season can also spawn great disasters.

NASA

near 70,000 feet; when they are westerly, as in 1995, hurricanes tend to be more frequent. Another factor is Caribbean surface pressures, which were the lowest ones in 50 years. In addition, winds at 40,000 feet over the tropics did not oppose surface easterlies. Finally, warm tropical waters in the Pacific, associated with El Niño, dissipated before the season began. El Niño had persisted through the weak 1991-94 seasons.

Gray also is studying a potentially powerful oceanic variation, the conveyer belt of warm water that travels northward along the east Atlantic, is replaced by cooler saltier water that sinks, and then heads south from the Arctic Ocean. This circulation may enhance hurricanes when it speeds up, as it did in 1995.

Still, Gray's long-range forecasts don't identify specific landfalls. In 1995, an upper-level low pressure area clinging for months to the East Coast pulled most storms out into the Atlantic before they could do harm. "We were incredibly lucky," Gray says.

HURRICANES / FORECASTING

Forecasters at the National Hurricane Center got an indication of Hurricane Andrew's strength around 3:40 on the morning of its landfall in South Florida. Their building, then in Coral Gables, shuddered violently as a powerful gust knocked the two-ton radar off its perch on the roof. When the meteorologists emerged after the storm, they found cars flung atop one another in their parking lot.

The forecasters also use more sophisticated means to size up a hurricane: satellite imagery, airborne reconnaissance, and computer-model projections. First they get a fix on a storm's location and intensity. For many years this was nearly impossible unless a hurricane passed over land or a ship, but in April 1960 the first weather satellite spotted a storm near Australia, launching a new age in forecasting. Just over a year later, pictures from space revealed a telltale swirl over the Gulf of Mexico. Trouble was heading for Texas: Hurricane Carla eventually swept the coast with a 22-foot storm surge. Relatively few people died because more than 350,000 had fled in the greatest mass evacuation in American history to that time.

To help them judge hurricane growth, forecasters use satellites to watch for such significant developments as rotating bands, shearing clouds, and storm asymmetries. The intensity estimates derived from pictures taken from orbit can be deceiving, however. To get accurate readings of central pressure and peak winds, direct observations are necessary, and the U.S. Air Force Reserve and National Oceanic and Atmospheric Administration regularly fly through storms in aircraft equipped with radar. In 1996 plane crews began dropping instrument

PAUL CHESLEY

packets called dropwindsondes into storms to retrieve wind, temperature, pressure, and moisture profiles, thus increasing the quality of forecasts by 15- to 30-percent.

Detailed data from dropwindsondes are fed into simulations run on the Weather Service's supercomputer in Camp Springs, Maryland. Some models compare a current hurricane with previous hurricanes, while others use physical laws to approximate the atmosphere steering the storm and calculate it ahead in time.

The models don't always agree. In 1994, while Hurricane Gordon was over the Gulf of Mexico, most models predicted it would continue westward. But an experimental model with a sophisticated mathematical approximation of the storm's vortex predicted Gordon would turn sharply to the east and hit Florida. That's what it did. The successful model, which averages a phenomenal 20 percent fewer errors

Hurricane specialists (opposite) ready themselves for a marathon, stomach-twisting flight into the eye of a storm. Getting an accurate fix on a storm's strength and position pays off in timely warnings. The 21-hour advance warning possible for Hurricane Andrew enabled residents to flee the vulnerable Florida Keys (below), where hundreds of people died in the waves of a 1935 killer hurricane.

than its predecessors, joined regular forecasting computer runs the next year. Such improvements are critical for believable warnings that give enough time to evacuate shorelines.

Still, forecasters have to hedge their warnings because hurricanes are notoriously capricious. For instance, when two hurricanes get within about a thousand miles of each other, they can dance around a common center—an interaction called the Fujiwara Effect. Typhoons Doug and Ellie embraced this way in 1994. Doug made landfall at Shanghai, retreated out over the Yellow Sea, and wheeled back to blast Shanghai again. Other storms may meander all by themselves. Early in its development in 1979, Supertyphoon Tip wandered for two days in the Pacific. Research suggests the storm, which later became the strongest tropical cyclone on record, was suffering the effects of a tilt in its own structure.

Intensity changes are even harder to predict. Tropical Storm Gordon sputtered in cool November air and shearing winds, but then roared to life over warm water near Key West and in the Gulf Stream. Other storms look threatening but may hold their true fire. In 1992, Typhoon Omar's outflow was temporarily plugged up by nearby Tropical Storm Polly. Omar caused extensive damage when it struck Guam, but developed into a supertyphoon soon after escaping Polly's exhaust.

The intensification delay was little consolation to the forecasters responsible for the western Pacific. The Joint Typhoon Warning Center in Guam was knocked out of commission for a week by Omar, making 1992 an unusual, red-letter year for hurricane forecasters.

HURRICANES / STORM SURGE

The hurricane appeared to be heading harmlessly toward a slow death over the open Atlantic. At least that's what forecasters expected when it moved northward near Cape Hatteras, North Carolina, on the morning of September 21, 1938. But instead the hurricane sensed a gap between upper-air pressure systems and accelerated to 60 miles an hour, slamming into Long Island at 3 p.m.

A major hurricane had not hit there since 1815. When Long Islanders looked south and saw a margin of gray on the horizon, they thought it was a fog bank. Moments later they realized their fatal mistake. Ten-foot waves atop a 20-foot storm surge came rushing in.

The surge obliterated hundreds of homes, tossing inhabitants into the swirling waters. Barrier islands were swept so bare of buildings, roads, and vegetation that rescue workers resorted to phone company charts to determine where houses once stood. Powerful waves moved 50-ton boulders and cut a channel near Shinnecock where years of expensive dredging had proved futile. Farther north, up to 12 feet of water swirled through downtown Providence, Rhode Island. Matinee moviegoers climbed into balconies to escape the flood; commuters drowned in the streets. Nearly 700 people died. The speed and brute force of the hurricane earned it the nickname, Long Island Express.

In hurricanes everywhere, storm surges are usually the most dangerous force. This is true in the United States, where surge accounted for most of the deaths in Galveston in 1900, as well as halfway around the world in the Andhra Pradesh region of India. There a cyclone surge killed 8,000 people in a few hours in November 1977, and another inundation killed a thousand more in 1996.

Evacuation is the best hope before an oncoming surge. More than two million people evacuated in the face of Hurricane Andrew. What they escaped, veterans working in the Florida Keys unfortunately faced in 1935. Their relief train arrived too late as the most powerful hurricane in American history covered the Keys with water. Among the terrible night's survivors were 70 men who clung to a heavy railway tanker car 20-feet tall while 25-foot waters threatened to drown them. Others awoke dazed in treetops. More than 400 perished.

On the ocean hurricanes are well-known as water-heaving storms. The 98-foot waves the *Queen Elizabeth 2* endured during Luis in 1995 were uncommon, but hurricanes easily produce 50-foot waves in deep water. A storm surge is a different, potentially more devastating force. It is a wholesale rise in sea level caused by wind and pressure effects on the surface of the ocean, and is evident only when waters pile up against shore, especially at landfall.

The precipitous drop in air pressure in the eye of a hurricane causes part of the surge by sucking the sea upward as if into a giant straw. Yet even a Category-5 hurricane will raise sea level only about a yard this way, a far cry from the 30-foot surge that a storm like this can produce at landfall.

Winds and geography create most of the surge height. The sustained onslaught of winds can pile water against the shore. Thus an approaching hurricane may drive the sea level up the beach hours before landfall. Meanwhile, far at sea, the converging spiral of hurricane

MIKE LACA/WEATHERSTOCK

winds drives water in toward the eye. Rather than piling higher, the water sinks and spreads out—until the storm approaches shore. Like a tsunami, which crosses the ocean as a low swell, the waves and mound of storm-surge water are amplified as they approach the continental shelf. The waters in the eye have no escape but up and out, and gradually the surge waters form a high mound or wall of water.

On top of this elevated sea, towering waves pound the shore. In addition, sudden wind shifts drive water back and forth to devastating effect. Where the hurricane spiral blows winds away from shore, water sinks back into the sea. When the eye passes, winds shift and blow onshore. The water sloshes back inland with even more force than possible under a steady wind. On the other hand, if waters surge inland first, their downward retreat is exacerbated by gravity and debris such as logs and cars.

The surge and waves on top of it build over time, scrunching together and piling up because of friction with the bottom. The slope of the seafloor thus determines how dangerous a particular hurricane can be. The steep sea bottom around Jamaica, for example, presents little opportunity for water to pile up. The storm surge, though powerful, is relatively low. On the wide, gently sloping continental shelf off Galveston, Biloxi, or Long Island, however, the water encounters much friction over many miles, producing surges more than 20 feet high.

These heights are additions to local tides. Thus, high tide is a particularly dangerous time for a landfall. An even worse time is syzygy, the alignment of sun and moon that produces the highest tides. Along the Eastern Seaboard, syzygy occurs during the harvest moon, magnifying the effects of such legendary landfalls as the Great Miami Hurricane of 1926 and the Long Island Express (when it encountered high tide in New England).

Many of these factors must have contributed to the record 40-foot surge recorded in a cyclone in Australia in 1899. The surge must have been a powerhouse, for every small increase in wave height yields big increases in destruction. Like the power of a storm's winds, the power of a crashing, weighty wall of water increases exponentially with height. A 20-foot wave has triple the damaging power of a 12-foot wave. This exponential factor looms large: Even a lowly 3-foot breaker rattles a house with 1,000 pounds of pressure per square foot.

A storm surge doesn't work with horizontal push alone, however. It lifts houses. Three women in Gulfport, Mississippi, were busy covering windows during Hurricane Camille when they saw water bubble up through the

As terrible Camille approached the coast in 1969, 25 "hurricane-watch" party-goers chose to ride out the storm at the Richelieu Apartments in Pass Christian, Mississippi (opposite, upper). Only one person, and an exposed foundation (lower), survived the 24-foot-high storm surge that swept in from the Gulf of Mexico and scoured the low-lying area.

baseboard "like from a mountain spring." Next a geyser burst through a hole where a floor heater had been: The house had been wrenched from its foundation and set adrift. The three survived by climbing onto tables as the water rose shoulder high. A couple in South Carolina had a similar experience in Hurricane Hugo. Knowing their house was a thousand feet from the surf, the couple decided to ride out the storm. They spent hours floating on a cabinet as the house popped off its piers. When water almost reached the ceiling, the couple stood on chairs to breathe.

A house adrift is at the mercy of the waves. "We were on a wall, then on the ceiling, and then on the floor again," recalled one Rhode Island woman after her house rolled in the 1938 hurricane surge. Sometimes, though, the storm surge is a gentle giant, setting a house down with the contents undisturbed.

Storm surges are not limited to beaches. A river-borne tidal bore caused the sudden flooding in Providence during the 1938 hurricane. When a similar surge raced up the Chesapeake Bay during a hurricane in 1933, its waters swept onto bridges over the Anacostia River in Washington, D.C., killing ten people when a train was washed off its track. During an 1893 hurricane that probably killed 2,000 people on the Sea Islands of South Carolina, a 12-year-old boy fell overboard from a dredge near Port

way, rising seven feet above normal. A more potent surge along Florida's Lake Okeechobee in 1928 emptied one end of the lake while spilling through weak earthen dams at the other. Nearly 2,000 people died.

Despite the complexities of surges, scientists can model them with computers. The results are nightmarish scenarios for cities like New Orleans, much of which lies below sea level. For several reasons, New York City is also highly vulnerable. The hurricanes that strike it often move very fast, weakening little from cooler conditions. The speed drives wind and waves onto the shore of the New York Bight, which forms a right angle to a hurricane heading north. Winds ahead of the eye pile waters westward alongshore toward the mouth of New York Harbor. Numerous canals and rivers carry this water far inland.

Using these factors, Nicholas Coch of Queens University has shown that a Category-3 landfall could produce an astounding 26-foot storm surge in neighborhoods along Jamaica Bay. Water levels (not including wave peaks) would rise 24 feet near Kennedy Airport and Howard Beach and 20 feet in the harbor.

Had Gloria hit the Northeast a few dozen miles farther west in 1986, Coch warns, the area would have suffered a catastrophe. The damage would have been as unimaginable as the storm that caught New Yorkers by surprise in 1938.

Royal. He was found the next day on a riverbank 20 miles inland, unconscious but alive, saved by his life preserver.

Occasionally hurricane winds produce dangerous surges in inland lakes as well. Remains of Opal in 1995 caused a standing wave, or seiche, in Lake Erie. The lake first sloshed one way, dropping more than three feet along the New York shore, and then sloshed the other

HURRICANES / WINDS

A broad, neatly spiraling circulation with 145-mile-an-hour winds is destructive enough, but a closer look at a hurricane's winds yields many dangerous surprises. During Hurricane Andrew, in particular, the devil was in the details.

Hurricanes have long been known to spawn tornadoes inside the broad wind spiral. At landfall, most of them occur in the front right quadrant of the storm (with respect to the eye path) and are not necessarily common in the eyewall, where the hurricane's circulation is strongest. In 1967, Hurricane Beulah set a record with 115 tornadoes in Texas when it turned west along the shore, sending its strong surface easterlies beneath the continent's prevailing westerlies.

Andrew spawned tornadoes in Louisiana, but in South Florida it produced even worse winds without a tornado. The severity of wind damage prompted careful surveys by experts such as T. T. Fujita, the University of Chicago professor who sifted through the rubble and examined 2,000 aerial photos. Like William Redfield 170 years before, Fujita reconstructed the storm's circulation in part by noting tree-fall patterns. He found that F3 damage northwest of Homestead Air Force Base occurred after the eye had passed, indicating that the storm had not weakened immediately after landfall. Surviving homes stood next to flattened homes, indicating whirls as little as 50 feet across with 200-mile-an-hour winds. Starburst damage patterns showed ferocious downbursts from concentrated precipitation shafts.

The storm had a peculiarity that helped explain these oddities. First, Andrew's eyewall lacked symmetry at landfall. The curvature of the eastern side was smaller, forcing winds to whirl faster. Second, winds turning inward due to friction at the surface may have temporarily enhanced the collection of moist fuel that otherwise was slowly shut down by landfall.

Weather Service radars revealed amazing convection structures within the eyewall that probably spawned the F3 winds. The difference between the northern and southern eyewall winds (with respect to the ground) ranged up to 55 miles an hour, far more than could be explained by the storm's forward movement. At the spot where the northern eyewall moved over the coastline, a series of 7 convective cells spun up over about 40 minutes. The updrafts then stretched and raised vertical vortices much the way mesocyclones are rolled, stretched, and raised in supercells over land. The effect was profound: On the edge of the northern eyewall, the pressure was probably lower than at the center. And one by one the cells, with extremely heavy precipitation and high winds, rotated counterclockwise around the eye.

These thunderstorms within a hurricane were not a first, but they had never been documented so clearly over land. In Typhoon Ida over the Pacific in 1958, high-flying U-2 aircraft photographed possibly similar structures of vortices embedded in the eyewall. Weaker hurricanes have also grown separate supercells or mesoscale convective complexes—like those over the Great Plains—within their circulations. These features produce stronger updrafts than most hurricanes sustain.

It was not surprising then that a hurricane-reconnaissance crew flying through the turbulence over Miami during the landfall felt it was

Winds, not heavy rains or a storm surge, caused most of the damage during Hurricane Andrew in 1992. The patterns of destruction at South Florida's Homestead Air Force Base (below) helped wind expert T. T. Fujita propose an explanation: Small whirls of tornado-strength winds embedded in the hurricane demolished some buildings, while leaving neighboring structures virtually unscathed.

JOHN J. LOPINOT/PALM BEACH POST-TIMES

one of the bumpiest rides they had ever taken. Perhaps it reminded them of an even more harrowing flight researchers had in Hurricane Hugo. The plane nearly crashed into the Atlantic when an engine failed in severe turbulence. Lumbering under a full load of fuel, the P-3 staggered into the eye and circled for hours within the canyon of cloud until it was light enough to head home. The shaken scientists had encountered a whirl of winds within the 155-mph eyewall circulation. The small whirl, with a 220-mph peak, raced around the eye 5 times in just an hour and a half, producing a series of cells similar to those Andrew spun up at landfall. It was one surprise the researchers could have done without.

HURRICANES / BANGLADESH

"It seemed as if the Day of Judgment had come," said a survivor of the April 29, 1991, cyclone in Bangladesh, the worst natural disaster of the last quarter century.

Another man remembered dark clouds and the green sea. The southern sky glowed red and the wind overwhelmed all other sounds. Tying his children to wooden planks and himself and his wife to another, the man awaited the dreaded surge. Suddenly, the thatched roof collapsed, knocking him unconscious. When he awoke, his family lay safely on sand two miles away.

Others spent their last moments amid salt spray, driving rain, hot 145-mph winds, and churning waters. A man clutching a water urn was washed to the safety of a treetop nearly 8 miles from home. Some people died from the bites of salt-crazed snakes. Hundreds who were huddled in a storm shelter died when 20 feet of water charged through the door. On the island of Sonadia, however, the 650 who made it to shelter survived; on Hatia, 2,000 disappeared from land outside the island's protective dikes.

In all, 139,000 people died as a 20-foot tide washed far inland. The survivors, who packed shelters to 10 times capacity, emerged into a landscape littered with bodies and ruined crops. More than 800,000 homes were destroyed or had disappeared; nearly a million more were damaged. Perhaps half a million farm animals died. In some regions so much water remained that helicopters couldn't find enough dry land to set their relief parcels down.

Coastal Bangladesh, a delta of great rivers laden with silt, is fertile enough to support a teeming population of farmers, shrimpers, salt cultivators, and fishermen—more than 1,500

High above the Bay of Bengal, a satellite spots a killer cyclone tightly coiled and bearing down on Bangladesh (right). Despite its early detection, the April 1991 storm took 139,000 lives. Saltwater inundation by a 1985 cyclone left survivors (opposite) with little potable water, few salvageable crops, and the awful certainty that the future will bring more storms and more sorrows.

EARTH SATELLITE CORP.

people per square mile. Many of the landless poor stake claims on temporary islands remade each year by monsoon flooding.

The powerful tropical cyclones that descend on this region form some of the worst storm surges on earth. The narrowing head of the bay often funnels storm waves onshore. And the wide continental shelf is only 30 feet deep more than 60 miles out—perfect for building surges. Even normal tides sweep far inland.

A 1737 cyclone killed 300,000 people near Calcutta, sinking 20,000 boats. Another in 1789 destroyed Coringa. Some accounts claim that less than two dozen of the town's 20,000 inhabitants survived a succession of 3 waves. In 1876, perhaps 200,000 died in a wall of water reportedly 40 feet high in the Chittagong area that was struck in 1991.

The 1991 storm was the sixth cyclone since 1960 to kill at least 10,000 in Bangladesh. The worst, on November 12, 1970, hurled winds of

125 miles an hour onto shore. Waters rose more than 20 feet in a 15-hour onslaught on Chittagong, Noakhali, Patuakhali, and other districts. At least 300,000 people died. The storm helped trigger the revolution that brought independence from Pakistan.

With weather satellites monitoring the bay, such disasters come with warning, but many people do not leave. The 1991 storm arose off the Andaman Islands as a tropical disturbance and became a cyclonic storm on the 25th of April, 900 miles away from Bangladesh. In the following days it dawdled over the Bay of Bengal. By the 28th, it was a full-blown tropical cyclone. The coil of boiling cloud tightened and slowed before turning toward the coast. It sped up on the 29th as waves raced ahead of the eye.

Forecasters in the Bangladesh meteorological headquarters in Dhaka watched the cloudy spiral head northeast. They issued warnings 30 hours ahead, and again 15 hours ahead of landfall. But no preparation is sufficient for a storm surge—except getting out of its way. Volunteers spread news of the storm and urged residents to leave; they probably saved hundreds of thousands of people. Many people, however, accustomed to frequent, weaker cyclones, balked at walking to the distant shelters.

At 3 a.m. on the 30th, the meteorologists back in Dhaka waited for reports from the landfall. One veteran of the horrible night in November 1970 quietly studied the satellite images showing the fresh sorrows from the bay. All he could do was cry.

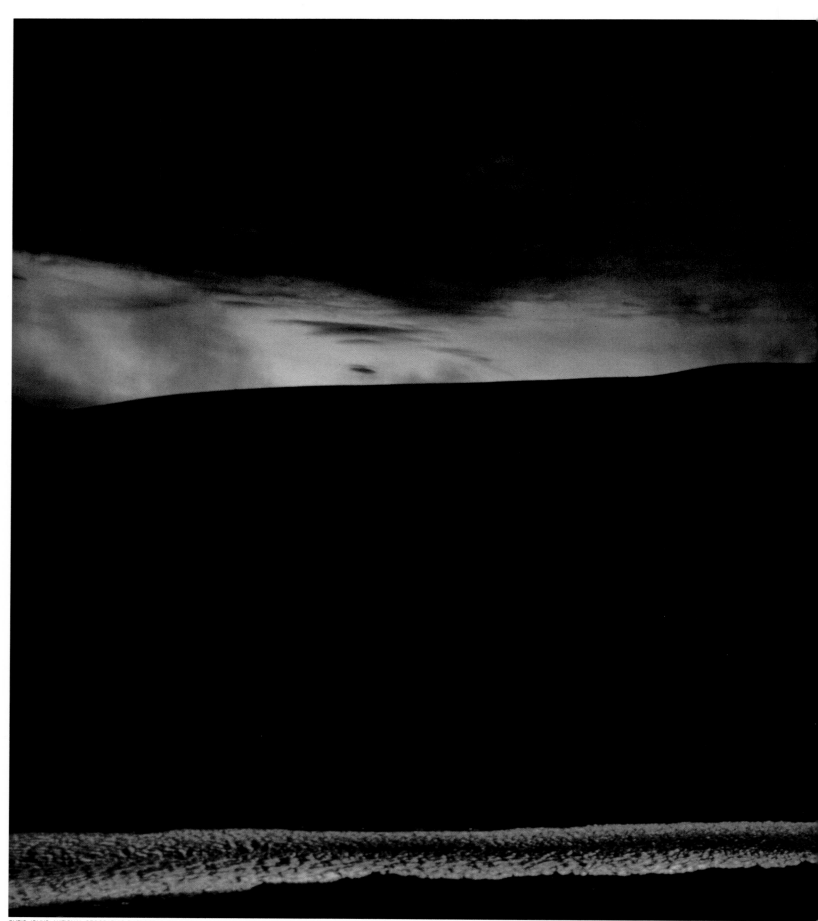

UNSTABLE LANDS
THE TERROR OF
TEMBLORS AND VOLCANOES

by Stephen L. Harris

INTRODUCTION

In Greek mythology, Zeus, king of the Olympian gods, defeated a giant dragon, Typhoeus, and confined him under Mount Etna, Europe's highest and most active volcano. Whenever Typhoeus awoke and struggled to break free from his subterranean prison, the earth shook and the mountain erupted streams of lava, rock melted by the monster's fiery breath. An embodiment of geologic violence, Typhoeus represented the chaotic natural forces that disrupt the human environment.

Although myth paints Typhoeus, the primordial serpent of chaos, as a threat to civilized order, he is actually an indispensable actor in the drama of the earth's age-long development. Unlike the moon—an inert rocky sphere lacking internal heat—the earth is a vibrant planet. When the ground lurches, damaging or destroying whole cities, or volcanoes erupt clouds of tephra that devastate towns, forests, and farmlands, it is not a mythical malignancy at work, but evidence that the planet is dynamic.

No matter how frightening they are to human beings, Typhoeus's antics are part of a long process that, over eons, has made the earth inhabitable. The forces of plate tectonics and countless earthquakes have elevated Italy's Alps and California's Sierra Nevada to their present scenic heights. Volcanic eruptions in the early history of the planet emitted vast quantities of water vapor and other gases, helping to create the earth's oceans and atmosphere.

We pay a price for our planet's continuing vitality. When the ground suddenly moved beneath Northridge, California, in January 1994, relatively few lives were lost, but this moderate shock, with a moment-magnitude of 6.7, caused property damage exceeding 40 billion dollars.

A year to the day later, on the opposite side of the Pacific Ocean, the port city of Kobe, Japan, was shaken by a similar quake. This time the cost in human terms was far larger: More than 5,000 people were killed, 25,000 injured, and about 300,000 left homeless. Damage amounted to a staggering 200 billion dollars. Even Kobe's ruin pales in comparison to other large 20th-century earthquakes, such as one that struck China in 1976, claiming an estimated 655,000 lives. Additional shocks, from California to Japan and from South America to China, are inevitable. Some will exact even higher costs in lives and property.

In recent decades, volcanic eruptions have also taken a high toll of lives. In June 1991, the Philippines's Mount Pinatubo, after a sleep lasting 600 years, produced one of the most violently explosive eruptions of this century. Fortunately, scientists monitoring Pinatubo's pre-eruption signals successfully persuaded authorities to evacuate tens of thousands of people living near the volcano, greatly reducing the number of casualties.

In 1985 geologists were also aware of the impending danger from Nevado del Ruiz, a volcano in Colombia, but nearby populations were not evacuated. When the volcano suddenly ejected a relatively small quantity of hot rock onto its summit ice cap, the resultant meltwater first generated a *jökulhlaup*, or glacier-outburst flood, which quickly turned into a large-volume lahar—a turbulent mixture of water-saturated rock debris—that rapidly streamed down the volcano's slopes and inundated adjacent valley

Bridges and overpasses suffer heavily in major earthquakes. In 1994, cars lie crushed under a collapsed section of the Golden State Freeway in Los Angeles, California. Striking before dawn on January 17, the moment-magnitude 6.7 Northridge earthquake blocked several crucial transportation routes.

PRECEDING PAGES: Lava fountains feed a river of molten rock pouring from Zaire's Nyamuragira volcano.

floors. In the town of Armero, an estimated 20,000 persons were buried alive when tons of volcanic debris resembling wet concrete swept through the area. The total number of fatalities reached 23,000, making the event the worst volcanic disaster since 1902, when a suffocating cloud of ash roared down the slopes of Mount Pelée, on the Caribbean island of Martinique, and snuffed out 30,000 lives.

As the world's population grows larger, more people will inevitably experience the terrifying sensation of feeling the earth buckle and twist beneath their feet. From ancient times through the present, people everywhere have felt an instinctive horror when terra firma has betrayed them. As the Roman philosopher Seneca asked after an earthquake devastated the Bay of Naples region in A.D. 62, "Yet can anything seem adequately safe to anyone if the world itself is shaken, and its most solid parts collapse? Where will our fears finally be at rest if the one thing which is immovable in the universe and fixed, so as to support everything that leans upon it, starts to waver?" If the earth loses its most reliable characteristic, stability, Seneca wondered, on what can we place our trust?

When long-dormant volcanoes such as Colombia's Nevado del Ruiz or Italy's Mount Vesuvius reawaken, millions of people must deal with Typhoean violence. The more we know about where earthquakes typically occur, what causes them, and why some volcanoes are more dangerously explosive than others, the more successfully we can identify potentially hazardous areas and take effective measures to minimize future losses of life and property.

PLATE TECTONICS

Earthquakes and volcanoes are world-wide phenomena, but they do not occur randomly all over the globe. Most of them are concentrated near plate boundaries, such as the Pacific Ring of Fire. A majority of the world's earthquakes take place in this zone, and about 70 percent of 1,511 potentially active land volcanoes are located there. Approximately 80 percent of the world's total volcanic activity is associated with mid-ocean ridges, submarine chains that extend around the earth as active zones of rifting and volcanism.

These linear belts mark boundaries between huge tectonic plates, fragments of the earth's crust that are always in motion. According to the theory of plate tectonics, the earth's crust is

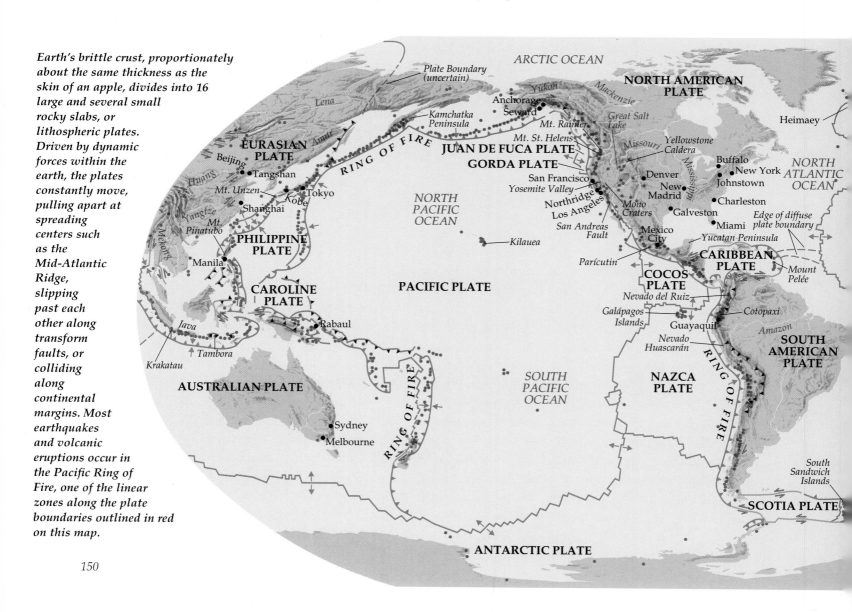

Earth's brittle crust, proportionately about the same thickness as the skin of an apple, divides into 16 large and several small rocky slabs, or lithospheric plates. Driven by dynamic forces within the earth, the plates constantly move, pulling apart at spreading centers such as the Mid-Atlantic Ridge, slipping past each other along transform faults, or colliding along continental margins. Most earthquakes and volcanic eruptions occur in the Pacific Ring of Fire, one of the linear zones along the plate boundaries outlined in red on this map.

150

composed of 16 major plates and several minor ones. These make up both the ocean basins and the continents. The crustal plates move slowly over the mantle, the 1,800-mile thick zone of semi-solid rock that lies between the earth's metallic core and its outer crust.

Forces within the earth drive the plates. For unknown reasons, the planet's internal heat is unevenly distributed, causing some parts of the interior to be hotter than others. The most intense concentrations of heat form convection cells, similar to ones in a pot of boiling water. This cycle of currents in the mantle, buoying the crust in one place and tugging it down in others, provides energy for plate movement.

Traveling at the speed of about two inches a year, the plates pull apart, collide, or grind past each other. The elongated zones at which plates separate are called divergent boundaries or spreading centers, where magma oozes up and creates new crust, gradually widening oceans and pushing continents apart.

Despite new material being added to its surface, our planet is not growing in size. Earth, in fact, destroys as much crust as it makes, recycling old rock along boundaries where oceanic and continental plates converge.

When plates bearing oceans collide with continental plates, the convergence typically forms a subduction zone in which dense, basaltic seafloor is thrust under relatively light granitic continental plates. As oceanic slabs descend into the mantle, they trigger quakes and build long ranges of coastal volcanoes.

When two continental plates converge, however, neither is subducted because both are relatively light and ride high on the mantle. Intense pressures along the zone of convergence may push the crustal margins skyward, eventually forming great mountain ranges.

In some places, adjacent plates form a strike-slip fault boundary, a region in which one plate slides horizontally past another one, sometimes imperceptibly and other times setting off large earthquakes.

ICELAND

Volga

Ob

EURASIAN PLATE

Charleville-Mézières

Irtysh

Mont Blanc

son-la-Romaine

Caspian Sea

Aral Sea

Vulcano

Vesuvius

Thera

Danube

Mediterranean Sea

Stromboli

Mt. Etna

AFRICAN PLATE

Nile

ARABIAN PLATE

Indus

Ganges

BANGLADESH

Bay of Bengal

INDIAN PLATE

Niger

S A H E L

Congo

East African Rift System

SOMALI PLATE

Nyamuragira

Longonot

Edge of diffuse plate boundary

INDIAN OCEAN

Zambezi R.

AUSTRALIAN PLATE

MID-ATLANTIC RIDGE

SOUTH ATLANTIC OCEAN

Strike-slip fault

Thrust fault

Volcano

Subduction zone

Spreading zone

PLATE MOVEMENT

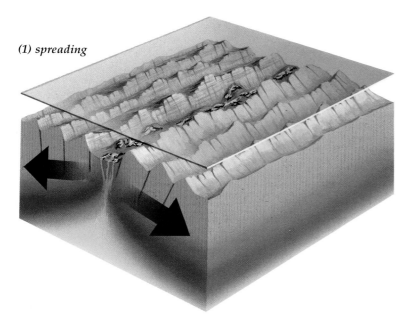

(1) spreading

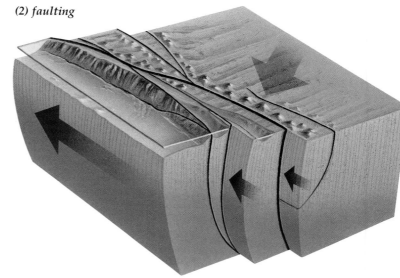

(2) faulting

One of the best known spreading centers, the Mid-Atlantic Ridge, began to form about 200 million years ago, a period when all the continents were joined together in a colossal landmass called Pangaea. This landmass started to break up when a great rift, a linear zone of deep fractures in the crust, gradually split the supercontinent. As plates on opposite sides of the rift pulled away from each other, forced apart by upwelling magma, the basin of the Atlantic Ocean began to form. Submarine eruptions continue today, causing the floor of the Atlantic Ocean to expand and pushing the Americas ever westward.

Although most spreading centers are hidden beneath oceans, the rifting process is clearly visible in Africa. The same forces that opened the Atlantic Ocean split the Arabian plate from the rest of the continent, forming the Red Sea. A new spreading center may be developing within the East African Rift system, site of numerous earthquakes and several active volcanoes. When the crust stretches and thins, tension cracks riddle crustal rock, which is then invaded by magma. Even when the magma does not reach the surface to cause volcanic eruptions, it increases pressure on the crust, producing additional fractures and expanding the rift zone. If the East African Rift continues to grow and deepen, the resulting depression will eventually be flooded by the Indian Ocean, transforming Africa's easternmost corner into a large island.

When ocean-bearing plates and continental plates converge, or collide, subduction zones usually develop. Here the heavy oceanic crust sinks under the edges of continents. Such zones surround much of the Pacific basin and are responsible for triggering thousands of earthquakes each year in Japan, Indonesia, western South America, Mexico, and Alaska.

Subduction also creates the extensive chains of volcanic mountains that encircle the Pacific.

Plate motion shapes the earth's surface. (1) Magma rising through crustal fractures builds mid-ocean ridges at spreading centers. (2) Plates slip past each other along transform faults. (3) Subducting ocean floor generates magma that fuels volcanoes. (4) Continents expand as converging plates add crustal debris to their margins.

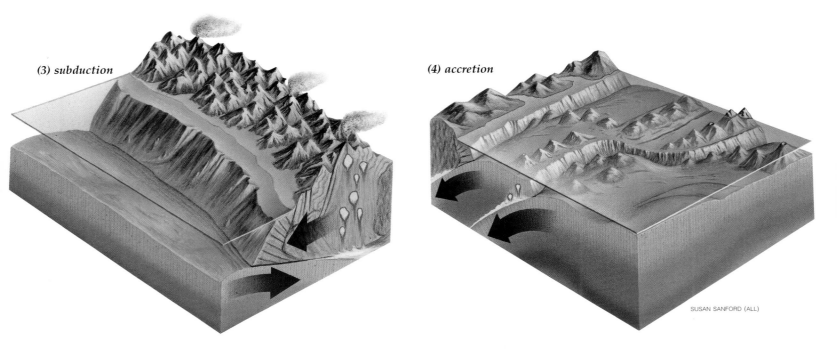

(3) subduction

(4) accretion

SUSAN SANFORD (ALL)

As water-saturated ocean floor pushes deeper into the mantle, the descending rocky slab encounters increasingly high temperatures. The addition of water to subterranean hot rocks lowers their melting point and generates new magma. Hotter and lighter than the surrounding rock, the magma rises into the crust, typically forming underground reservoirs of molten rock called magma chambers. When the chemically evolving magma becomes more buoyant than the surrounding rock and migrates to the earth's surface, a new volcano is born or an older one grows larger.

When continental plates collide, neither subducts. For example, when the plate carrying India ran into Asia about 50 million years ago, the southern border of the Eurasian plate partly overrode India, crumpling and folding crustal rock to form the Himalaya mountain range. Sometimes seamounts and other crustal material collide with continental plates and fuse with them, enlarging their margins in a process known as accretion.

Plates slide past each other at transform-fault boundaries, such as the one formed by California's San Andreas Fault system. A complex zone of parallel branching faults up to 60 miles wide, the system extends northward from the Mexican border to Cape Mendocino, defining the boundary between this part of western North America and the eastern Pacific plate.

Moving slowly to the northwest, the Pacific plate carries a sliver of California on its margin, including the Los Angeles region, which will become a suburb of San Francisco in a few million years. The Los Angeles City Hall is already ten feet closer to the bay area than when it was built in 1924. Eventually this segment will detach itself from the rest of California and head north to Alaska. In the meantime, as the plates slide past each other, the San Andreas fault system generates some of the West's largest quakes.

153

EARTHQUAKES

An earthquake is the vibration or shaking of the ground caused by a sudden movement or rupture of large masses of rock within the earth's crust or upper mantle. Most earthquakes originate within the top 10 to 20 miles of the mantle and crust, the so-called lithosphere. Powerful forces in the lithosphere, such as plate movement, exert stress on crustal rock. Rock is elastic enough to accumulate tremendous amounts of strain, bending or changing shape and volume. When stress exceeds the strength of the rock, as when a rubber band is stretched and snaps, the rock breaks along a preexisting or new fracture plane called a fault. The fracture rapidly extends outward from its place of origin, called the focus or

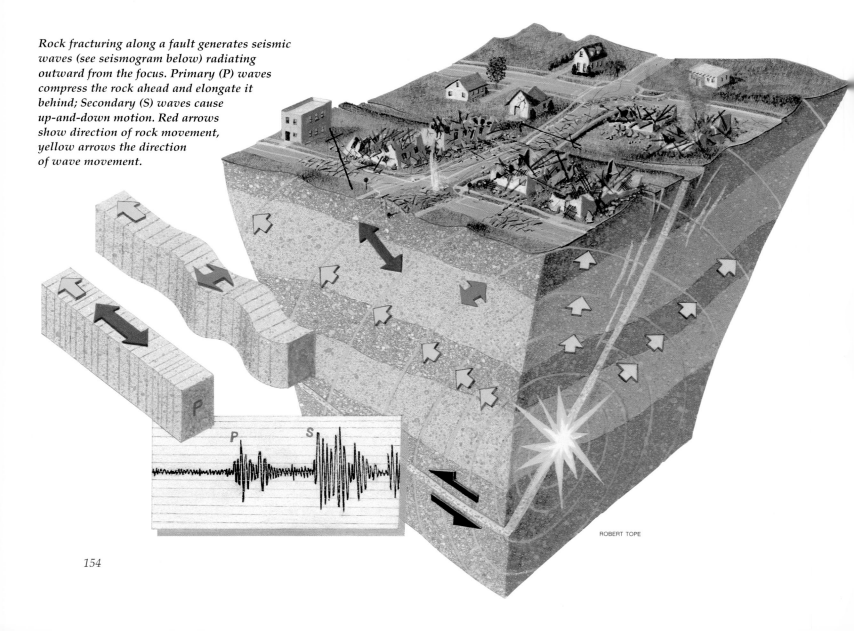

Rock fracturing along a fault generates seismic waves (see seismogram below) radiating outward from the focus. Primary (P) waves compress the rock ahead and elongate it behind; Secondary (S) waves cause up-and-down motion. Red arrows show direction of rock movement, yellow arrows the direction of wave movement.

ROBERT TOPE

154

hypocenter. As rock breaks, waves of energy—seismic waves—radiate through the earth and over the surface, causing an earthquake. The point on the surface directly above the focus is called the epicenter, the place where ground shaking is typically most severe.

Seismic Waves

The friction and crushing caused by masses of bedrock sliding past one another along a fault generate two basic kinds of seismic waves: Body waves and Surface waves.

Body waves spread out in every direction, moving swiftly through the earth's interior. These waves travel to distant points on the earth's surface and are subdivided into Primary (P) waves and Secondary (S) waves. The P waves (also called compressional waves) move the fastest: They compress the rock in front of them and elongate it behind as they rush through the planet at speeds of three to four miles per second. Next come the S (also called shear) waves, which undulate, causing an up-and-down and side-to-side motion as they roll through the earth at about two miles per second.

Most damaging to human-built structures are Surface waves, the second basic type, which travel on the earth's surface away from the epicenter much like ripples on a pond. Surface waves are divided into Rayleigh (R) waves and Love (Q) waves. Shaking the ground both vertically (R waves) and horizontally (Q waves), they can create a visible rolling and billowing of the surface as well as a jerky zigzag motion that is destructive to high-rise buildings. The slowest moving of all seismic waves, Surface waves cause the worst devastation. They accelerate ground motion and take longer to travel through a given area. During a 1992 quake along California's north coast, an observer described Surface waves as violently jerking the ground "up and down like a piston."

Seismographs and Seismograms

The vibrations that earthquakes produce are detected and recorded by instruments called seismographs. When the earth shakes, a seismograph records the ground motion as a sequence of sinusoidal waves expressed as a tight series of sharp peaks on a seismogram. The pattern of peaks along a line trace reflects the varying intensity of the vibrations as they are received.

An international network of seismograph stations is maintained all over the planet, so that within just a few minutes of a large earthquake seismographs at many widely scattered locations record the seismic waves it produces.

Because the different kinds of seismic waves travel at different speeds, they are picked up by a seismograph in an established order. Although Body waves and Surface waves originate at the earthquake focus (hypocenter) at the same time, the P waves arrive first, the S waves next, and the Surface waves third. The time intervals between the first arrivals of the P, S, and Surface waves increase with distance from the focus.

By analyzing the time lapse between consecutive arrivals of Body and Surface waves, earth scientists can quickly calculate the approximate distance from a seismograph station to the source of the earthquake. When comparing the records at three or more stations, seismologists can also determine the earthquake's location anywhere in the world.

EARTHQUAKES / MEASUREMENTS

Seismologists have devised several methods for determining the severity of an earthquake. The Richter scale, which Charles F. Richter and Beno Gutenberg developed in the 1930s and '40s to measure the relative strength of quakes in southern California, is commonly used to calculate the magnitude, or amount of energy a given earthquake releases. In recent years seismologists have designed a new scale, called moment-magnitude, for measuring exceptionally large earthquakes. This method, which has become increasingly accepted in the scientific community, computes an earthquake's energy release by a formula involving the surface area of the fault rupture and the amount of rock displacement along the fault. Yet another scale, the Modified Mercalli Intensity scale, measures an earthquake's intensity, or effects, at a specific location.

The Richter Scale

A numerical scale that assigns numbers to earthquakes in ascending order of magnitude, from zero to slightly less than 9, the Richter scale is logarithmic. When earthquake magnitude increases by a whole number, such as from 5 to 6, it means that the strain energy released is ten times greater.

The size of seismic waves, which determines the rate of ground shaking, is commonly expressed as amplitude. This term refers to the maximum height of a seismic wave crest or the depth of its trough as recorded on a seismogram. Earth scientists estimate that a tenfold increase in seismic wave amplitude signifies an increase of about 32 times in energy release. Thus, the Richter-magnitude 7.1 Loma Prieta quake that shook the San Francisco Bay Area in October 1989 was many times more powerful than the 6.4 shock that ravaged Northridge, in southern California, 5 years later. The great San Francisco earthquake of 1906, Richter magnitude 8.3, was hundreds of times stronger. This logarithmic increase, however, does not mean that a magnitude 6 quake will shake your house a thousand times harder than a magnitude 4. The former earthquake will simply last longer and release its tremendous amount of energy over a much larger area.

Richter magnitudes are also used to classify earthquakes. Quakes that register about 2 or less are called microearthquakes and are usually not perceptible to human observers. An earthquake measuring about 6 is considered "moderate," though such a quake is certainly capable of inflicting considerable damage when centered in a heavily populated area. A magnitude 7.1 event, such as the Loma Prieta earthquake or the one that shook Washington State's Puget Sound region in 1949, is categorized as "major." If a quake has a Richter reading of 8 or above, it releases a huge amount of strain energy, attaining the status of a "great" earthquake.

The magnitude of a particular event does not, however, necessarily determine its destructiveness. Unless it generates a series of large sea waves called tsunamis, a great quake originating far out on one of the earth's oceanic plates may not even be noticed onshore. But a comparatively moderate quake centered near a densely populated area, such as the magnitude 6.4 event that hit Northridge in 1994 or the 6.8 quake that ravaged Kobe, Japan, in 1995, can inflict extensive damage.

MODIFIED MERCALLI INTENSITY SCALE

■ I Rarely felt by people.

■ II Felt indoors only by persons at rest. Some hanging objects may swing.

■ III Felt indoors by several people. Hanging objects may swing. Vibration like passing of light trucks. Duration estimated. May not be recognized as a quake.

■ IV Felt indoors by many people, outdoors by few. Hanging objects swing. Vibration like passing of heavy trucks; or sensation of a jolt like heavy ball striking walls. Standing automobiles rock. Windows, dishes, doors rattle. Wooden walls and frames may creak.

■ V Felt indoors and outdoors by nearly everyone; direction estimated. Sleepers wakened. Liquids disturbed, some spilled. Small unstable objects displaced or upset; some dishes and glassware broken. Doors swing; shutters, pictures move. Pendulum clocks stop, start, change rate. Swaying of tall trees and poles sometimes noticed.

■ VI Felt by all. Damage slight. Many frightened and run outdoors. Persons walk unsteadily. Windows, dishes, glassware broken. Knickknacks and books fall off shelves; pictures off walls. Furniture moved or overturned. Weak plaster and masonry cracked.

■ VII Difficult to stand. Damage negligible in buildings of good design and construction; slight to moderate in well-built buildings; considerable in badly designed or poorly built buildings. Noticed by drivers of automobiles. Hanging objects quiver. Furniture broken. Weak chimneys broken. Damage to masonry; fall of plaster, loose bricks, stones, tiles, and unbraced parapets. Small slides and caving in along sand or gravel banks. Large bells ring.

■ VIII People frightened. Damage slight in specially designed structures; considerable in ordinary substantial buildings, partial collapse; great in poorly built structures. Steering of automobiles affected. Damage to or partial collapse of some masonry and stucco. Failure of some chimneys, factory stacks, monuments, towers, elevated tanks. Frame houses moved off foundations if not bolted down; loose panel walls thrown out. Decayed pilings broken off. Branches broken from trees. Changes in flow or temperature of springs and wells. Cracks in wet ground and on steep slopes.

■ IX General panic. Damage considerable in specially designed structures; great in substantial buildings with some collapse. General damage to foundations; frame structures, if not bolted, shifted off foundations and thrown out of plumb. Serious damage to reservoirs. Underground pipes broken. Conspicuous cracks in ground; liquefaction of soil.

■ X Most masonry and frame structures destroyed. Some well-built wooden structures destroyed. Serious damage to dams, dikes, embankments. Landslides considerable. Water splashed onto banks of canals, rivers, lakes. Sand and mud shifted horizontally on beaches and flat land. Rails are bent slightly.

■ XI Few, if any, masonry structures remain standing. Bridges destroyed. Broad fissures in ground; earth slumps and landslides widespread. Underground pipelines completely out of service. Rails bent generally.

■ XII Damage nearly total. Waves seen on ground surfaces. Large masses of rock displaced. Lines of sight and level distorted. Objects thrown upward.

Moment-Magnitude Scale

To measure the energy released by the largest earthquakes, scientists have recently developed the moment-magnitude scale, which uses a formula based on rock displacement and surface area of the rupture. In general, the larger the break along the fault, the larger the size of the earthquake. Whereas the Richter scale calculates magnitude by comparing seismograph readings from stations all over the globe, the moment-magnitude scale incorporates field-work—the physical measurement by geologists of fault movement and areas affected by it—when assigning magnitude numbers to quakes.

MAJOR 20TH-CENTURY QUAKES

		Moment-Magnitude	Deaths
■ 1960	Southern Chile	9.5	5,700
■ 1964	Southern Alaska, U.S.	9.2	131
■ 1985	Michoacán, Mexico	8.1	9,500
■ 1923	Tokyo, Japan	7.9	142,810
■ 1906	San Francisco, U.S.	7.7	3,000
■ 1976	Tangshan, China	7.4	655,000
■ 1989	Loma Prieta, U.S.	7.0	62
■ 1995	Kobe, Japan	6.9	5,200
■ 1994	Northridge, U.S.	6.7	60

The most powerful earthquake ever recorded—moment-magnitude 9.5—struck the coast of Chile on May 18, 1960. The second largest event, and the strongest quake ever registered in North America, was a moment-magnitude 9.2 temblor that jolted a vast region in Alaska on March 27, 1964. Both of these events were subduction-zone earthquakes, and both produced tsunamis.

Measuring Intensity

Whereas a magnitude number expresses the overall size and strength of an earthquake, intensity refers to a quake's effects at a specific location. The great San Francisco earthquake of 1906, for example, was assigned a single Richter magnitude of 8.3, but it had a wide range of intensities in different places.

How violently the earth shakes in a given area depends not only on its distance from the quake's epicenter (a point on the earth's surface directly above the subsurface focus of the earthquake), but also on the geologic composition of the ground through which the seismic waves move. Surface waves passing through solid bedrock can initiate strong vibrations, but their effects are catastrophically magnified when they move through loose, unconsolidated soil. The most damaging effects typically occur at sites that were created by adding landfill to old marshes, lake beds or riverbeds, especially when the groundwater level is high. Amplitude (wave motion) increases significantly in waterlogged sediments, causing liquefaction.

Some of the worst destruction in the 1985 Mexico City earthquake occurred where an old lake basin, site of the original Aztec capital, had been filled in to provide new land for urban growth. In many places, a line of demarcation marking different degrees of soil liquefaction, and hence of structural damage, was clearly drawn. Outside the former lake shoreline, most large buildings stood relatively unscathed. A block or two away, many comparable structures, erected on unstable lake sediments, toppled as if they had been constructed of children's play blocks.

The enormous disparity between similar well-designed buildings that stood on different kinds of ground is clearly evident in most earthquakes striking urban areas. In both 1906 and 1989, the San Francisco Bay neighborhoods most heavily damaged were typically those

Major earthquakes, such as the moment-magnitude 8.1 quake that ravaged Mexico City in 1985, can destroy facilities needed to deal with the crisis. The collapse of the Benito Judrez Hospital caused floors to pancake, entombing about 1,000 patients and medical staff. About 9,500 persons died in the disaster, with the worst losses concentrated at the site of an ancient lake bed, where shaking caused liquefaction.

HERMAN KOKOJAN/BLACK STAR

built atop landfill, where liquefaction was common. The same was true for Kobe in 1995.

Modified Mercalli Intensity Scale

The most commonly used system to rate earthquake intensity is the Modified Mercalli Intensity scale, which distinguishes 12 degrees of severity. Ratings are based on direct observation and evaluation. For example, the maximum intensity during the 1964 Alaska quake was a X; the damage from the San Francisco and New Madrid quakes indicated a maximum of XI.

EARTHQUAKES / SAN FRANCISCO 1906

Perhaps the first person to see the 1906 earthquake hit town was a railroad mechanic taking a swim at Ocean Beach on San Francisco's western shore. Dawn was just breaking at 5:12 a.m., April 18, when Clarence Judson emerged from the Pacific surf, only to have the earth convulse beneath him. As Judson struggled to stand and run for his clothes, he watched seismic waves roll landward. "The motion of the quake was like the waves of the ocean—about twenty feet between crests—but they came swift and choppy," he recalled, "with a kind of grinding noise.…"

A few miles farther east, in downtown San Francisco, police sergeant Jesse Cook also witnessed the quake's arrival. Pausing on his morning rounds to chat with a produce worker, Cook suddenly heard a distant low rumbling, "deep and terrible." Looking up Washington Street, he saw the ground pitch and undulate "as if the waves of the ocean were coming toward me, and billowing as they came." For 45 to 60 seemingly interminable seconds, San Francisco was shaken by the most powerful earthquake ever to strike a major city in the continental United States.

The magnificent city hall, with its Greco-Roman columns and dome higher than that of the nation's Capitol in Washington, D.C., crumbled like dried frosting on an elaborate wedding cake. Broken columns and huge blocks of concrete crashed to earth as the largest building in the West lost its faux-marble veneer, leaving only twisted steel girders to support the huge dome. San Francisco's wrecked municipal center became a potent symbol of the city's ruin.

Masonry facades, cornices, and cemetery monuments tumbled throughout the city, but the quake was more deadly in some parts of town than others. Well-designed structures on bedrock, such as the mansions atop Nob Hill, survived almost unscathed, while poorly constructed brick and wood-frame edifices on landfill near San Francisco Bay suffered heavy damage.

Visitors staying at upscale hostelries, even when severely shaken, fared commensurately better than many transients in hundreds of low-cost hotels and boardinghouses. Guests at the luxurious Palace Hotel, one of the city's most ornate showplaces, included world-famous tenor Enrico Caruso and other artists with New York's Metropolitan Opera company, which was then performing in San Francisco. Rudely thrown from his bed, Caruso awoke to chaos: "Everything in the room was going round and round. The chandelier was trying to touch the ceiling, and the chairs were all chasing each other. Crash—crash—crash!… Everywhere the

Soldiers from the Presidio patrol Market Street on April 18, 1906, as fire eats its way toward San Francisco's 18-story Call Building. To the left, a banner flies gallantly above the Palace Hotel, where Enrico Caruso stayed during the devastating 8.3 Richter-magnitude earthquake. Horses killed by bricks from falling walls lie amid the rubble (opposite).

walls were falling and clouds of yellow dust were rising.... I thought it would never stop." Other guests noted the violent twisting sensation that gripped their hotel, emphasizing "the rotary motion of the immense building" that made it seem "to dance a jig."

The seven-story Palace Hotel, which occupied a full block on Market Street, rode out the quake battered but structurally intact. Its builder, William Ralston, had learned a lesson from the 1868 Hayward quake and spent a fortune on 3,000 tons of "earthquake-proof" iron banding to reenforce the Palace's thick brick walls. Other lodgings, in the crowded working-class district south of Market, were not so carefully designed, nor their guests so fortunate. The four-story Valencia Hotel, built atop a former marsh, sank into the earth, its top story brought level with the street. Dozens of its occupants drowned when broken water mains flooded telescoped lower floors. More deaths

occurred when a block of wood-frame buildings collapsed and caught fire. Frantic attempts to rescue people who were trapped in the ruins were largely unsuccessful. Eyewitnesses estimated that 150 to 300 persons were killed in the Brunswick House alone.

Official estimates limited the death toll to a few hundred, but recent studies indicate a far higher number of fatalities. Gladys Cox Hansen, archivist of the city and county of San Francisco, uncovered evidence that more than 3,000 people died in the quake and its aftermath.

The effects of the earthquake were felt far beyond San Francisco. When the San Andreas Fault broke on April 18, the northernmost section ruptured from near San Juan Bautista to Cape Mendocino, a distance of about 280 miles. Lurching alongside the North American plate, the Pacific plate moved horizontally 25 feet to the northwest at Point Reyes and up to 28 feet at Shelter Cove, visibly furrowing the surface along much of the fault line. By contrast, the southern segment of the fault break had an average displacement of only 9 to 12 feet.

Ground motion that shattered San Francisco also caused severe damage in other coastal cities and towns. The campus of Stanford University was particularly hard hit; massive Romanesque structures, including the library and memorial church, were virtually demolished. One hundred persons died when the poorly built Agnews Hospital for the insane almost totally collapsed. At points along the coast, steep cliffs plummeted into the sea, the earth split in gaping fissures, railroad engines and cars were thrown from their tracks, and rails were contorted into grotesque shapes. South of Santa Clara, landslides from both sides of the gulch in which the Loma Prieta sawmill was located completely buried the mill, entombing the nine men working there.

Only moments after the ground had ceased shaking in San Francisco, fires from overturned stoves, broken gas mains, and damaged electrical wiring broke out all over the city. Within 15 minutes after the quake, an estimated 50 fires were reported, most in the densely populated area south of Market Street. By noon on April 18, a square mile in the city's heart was aflame. During the next three days the urban conflagration raged virtually unchecked, creating a firestorm that ultimately consumed more than 28,000 buildings.

On Wednesday afternoon, the novelist Jack London watched San Francisco's destruction from the bay. "It was dead calm," he wrote, "Not a flicker of wind stirred. Yet from every side wind was pouring in upon the city.... The heated air rising made an enormous suck. Thus did the fire of itself build its own colossal chimney through the atmosphere. Day and night the dead calm continued, and yet, near to the flames the wind was often half a gale, so mighty was the suck." The phenomenon London described is typical of the firestorm, which generates a column of intensely hot air and burning gases that rises above the fire. As the fire consumes oxygen, it draws air inward from all sides to replace it. Eyewitnesses described the updraft thus created as tearing roofs from houses and carrying them skyward as if they were sparks. Fierce heat from the advancing conflagration caused whole buildings to explode in flame even before the fire itself reached them. Temperatures at the heart of this inferno approached 2,000°F, twisting and melting steel and reducing glass to shapeless blobs.

The earthquake had shattered many of the city's water mains, severely reducing the amount of water available to fight the fire. Although their chief, Dennis T. Sullivan, died from injuries incurred during the quake, the San Francisco fire department rallied quickly to wage a long and gallant battle, pumping water from the bay, sewers, and ditches that were flooded by broken pipes. Early attempts to create firebreaks by dynamiting buildings were almost entirely futile, commonly scattering burning embers and igniting more fires. It was not until a U.S. Navy team arrived on the scene the third day of the fire that there was effective use of dynamite.

When the last threatening blaze died about 7 a.m., Saturday, April 21, more than four square miles—including San Francisco's entire

With one of American history's largest urban fires rapidly approaching, throngs of San Franciscans walk or ride down Golden Gate Avenue to find refuge in the city's parks and open squares. Crumpled wood-frame structures typify hundreds of similar buildings that fell off foundations or collapsed when the earthquake churned the loose, wet soils on which they rested into the consistency of quicksand.

Wherever possible, all signs of earthquake damage were carefully erased from photographs released to the press. Thus retouched, the tower of the Hall of Justice, which toppled in a strong aftershock at 8:14 on April 18, was miraculously restored to the perpendicular in pictures that showed the approaching fire. The Ferry Building tower, a prominent landmark at the foot of Market Street, leaned precariously and displayed serious cracking in the stone facade, but the tower was airbrushed back into plumb for its post-earthquake appearance on postcards. The city hall was too spectacular a ruin to disguise with camera magic, but rumors circulated that it suffered, not from the incredibly intense shaking it actually experienced, but from shoddy construction—an inaccuracy that insurance inspectors later refuted.

business, financial, and most major residential districts—had been reduced to ashes. Approximately 250,000 people were homeless.

As local historians have recently discovered, the city was still smoldering when several large corporations began a campaign of disinformation about the disaster. Assuming that East Coast financiers would not invest capital in a region susceptible to earthquakes, executives of the Southern Pacific Company, which suffered enormous dollar losses to its railroad facilities, formulated an influential policy of minimizing the quake's impact. Accordingly, the 1906 catastrophe was officially designated "the fire" and the quake was virtually ignored.

Only a week after the earthquake, the *San Francisco Chronicle* published a resolution passed by the Real Estate Board: The phrase "the great earthquake" should be abandoned and replaced by "the great fire." Whereas quakes were strange and frightful events, major conflagrations—such as the ones that burned Chicago to the ground in 1871 and decimated Baltimore in 1904—were familiar and presumably manageable threats. As the late Herb Caen, a columnist for the *Chronicle,* wrote years later, the 1906 calamity became the world's only fire to register 8.3 on the Richter scale.

EARTHQUAKES / LOMA PRIETA

National attention focused on the San Francisco Bay area on October 17, 1989. The third game of the World Series between San Francisco and its rival across the bay, Oakland, was about to begin. Fans crowding Candlestick Park were awaiting the first pitch when, at 5:04 p.m., the stadium began to vibrate. Then, as electric power failed, television screens went blank, leaving audiences across the country guessing at what had gone wrong.

For 15 terrifying seconds, northern California was shaken by a moment-magnitude 7 earthquake. Soon after, dispersing baseball fans and rush-hour commuters found many streets and highways impassable. A column of smoke and flame expanded rapidly over San Francisco's Marina District, while other fires smoldered in the East Bay. As night deepened, the area lay in near-total darkness, punctuated only by car lights and the red flashes of emergency vehicles.

Operating on emergency power, local television and radio stations slowly gathered the news: A 50-foot section of the San Francisco-Oakland Bay Bridge's upper deck had collapsed. The Embarcadero Freeway near San Francisco's waterfront suffered heavy damage. Landslides and weakened bridges closed Highways 1 and 101 to the south. Sixty-two persons lay dead or dying, while another 3,800 incurred injuries requiring medical attention.

The worst loss of life occurred on Interstate 880 in Oakland, where 44 slabs of concrete deck, each weighing 600 tons, fell on vehicles below. Observers said the quake rolled along the two-decker freeway like a colossal ocean wave behind which section after section fell on the lower roadway, compressing cars to a thickness of 18 inches. Despite fears that aftershocks would bring down the structure, many citizens living nearby rescued drivers trapped in what had become a concrete tomb. Later, fire crews and other professionals labored for days to retrieve 41 bodies from the wreckage.

The earthquake, which damaged 24,000 houses and apartment buildings and about 4,000 business structures, originated south of San Francisco in the Santa Cruz Mountains. Named for a peak near its epicenter, the Loma Prieta earthquake was triggered when a 30-mile-long segment of the San Andreas Fault abruptly shifted, pushing the crustal block west of the fault northward by five or six feet. The focus lay about 11.5 miles underground and did not rupture the ground, although vigorous shaking opened numerous fissures over an area 30 miles long and 12 miles wide.

The largest surface crack, located in the Santa Cruz Mountains, measured 650 yards long and 2.5 yards across. Many mountainside homes here were pitched from their foundations or suffered total or near collapse when spreading, sagging, and cracking of the ground caused them to fail. The town of Santa Cruz, a few miles from the epicenter, was extremely hard hit. Three people were killed by falling walls in the Pacific Garden Mall, a six-block cluster of Victorian-era brick buildings.

The extent of damage 50 to 60 miles distant might seem surprising. Although seismic waves weakened as they approached the bay area, the ground shaking they produced was strongly amplified as they passed upward through water-saturated landfill along the margins of San Francisco Bay. Soil liquefaction in

ROGER RESSMEYER/CORBIS

the Marina District, which was largely built on rubble from the 1906 quake and fire, caused the collapse of 35 buildings and severe damage to about 150 others. Most of the buildings damaged in San Francisco stood on landfill sites that had once been sloughs, marshes, or streambeds.

The irregular distribution of damage indicates the importance of imposing strict codes for structures near a fault or on soft soils. Whereas downtown Santa Cruz, built atop old river sediments, lost many wood-frame and masonry structures, the nearby University of California at Santa Cruz, sitting on bedrock, survived with minor damage. Stanford University, which suffered heavily in 1906, experienced severe shaking that caused 165 million dollars in damage. The sprawling megalopolis of San Jose had only minor losses. Even in San Francisco and Oakland serious damage was confined largely to isolated pockets. In most neighborhoods, observers would be hard pressed to find visible evidence of a major earthquake.

EARTHQUAKES / NORTHRIDGE

The moment-magnitude 6.7 earthquake that struck Northridge, California, on January 17, 1994, sounded a seismic warning for the entire Los Angeles region. This suburb 21 miles northwest of downtown Los Angeles had been considered seismically well prepared. But the quake wrecked 12,500 buildings and 10 bridges, closing 3 major freeways. Economic losses exceeded 40 billion dollars, making it more costly than Hurricane Andrew.

The high costs resulted from two crucial factors: The epicenter lay close to a major city; the quake caused some of the strongest shaking ever recorded. The tremor originated on a blind thrust fault—the kind that can generate bone-jarring vertical motion. Such a fault is nearly horizontal to a gently sloping buried fault, one end of which slants at a shallow angle to the surface. When a thrust fault ruptures along a shallow focus, one side thrusts upward and the

Reducing many homes to rubble (left) in Northridge, California, a devastating quake on January 17, 1994, offered a dress rehearsal for the disaster that will occur when a comparable or larger jolt centers under downtown Los Angeles. Flames shoot from a ruptured gas line and water flows from burst mains near homes burned to the ground following the moment-magnitude 6.7 quake (opposite).

release of seismic strain energy so close to the surface causes severe ground motion.

Because the Northridge quake hit at 4:31 on a holiday morning, there were only 60 fatalities. Had it occurred at midday, casualties would have been higher, particularly in shopping centers and department stores, several of which suffered near-total collapse.

Inspectors described the damage to many modern reinforced concrete structures as generally "tremendous," and older masonry buildings were so damaged that they had to be torn down. The Northridge campus of California State University, located almost directly above the focus at the earthquake's epicenter, suffered heavy damage.

One lesson from the Northridge earthquake is that southern California faces many seismic threats. As an increasing number of previously unrecognized faults are discovered and mapped, geologists conclude that another major danger lies in a network of faults in the Los Angeles basin. The problem is magnified by what they call the "big bend." As the San Andreas Fault approaches Los Angeles from the south, it bends westward before resuming a more northerly direction along the coast. That bend, running 30 miles east of the city, places the North American plate in the path of southern California, which is riding the Pacific plate northward.

Given the enormous seismic strain in the Los Angeles basin, some geologists think this region has not had enough earthquakes in historic time. "We need 15 more Northridges to catch up with the strain that's built over the past 200 years," asserts Jim Dolan, a geologist at the University of Southern California. "Or else one big event. It could be even larger."

"We used to think that the main danger to L.A. was a big one originating on the San Andreas Fault, which runs at least 30 miles from downtown," says Kerry Sieh of Caltech. "But the San Andreas is no longer the only source for a big one. We now have the enemy right beneath the city as well as on its margins."

A 7 or 7.5 shock occurring on a thrust fault here may cause more damage than an 8 on the San Andreas. One prospective source of catastrophe is the Elysian Park thrust-and-fold system extending under downtown Los Angeles. Another is the Hollywood Fault, whose movement might cause the adjacent Santa Monica Fault to break, generating a magnitude 7+ shock that could decimate an area from east of downtown through Beverly Hills to Malibu. A quake on the Newport-Inglewood Fault could cause even more devastation. From the ocean floor south of Laguna Beach, the fault extends northwestward to Beverly Hills. If it slipped an average of 3 feet along a 45-mile-long break, the ground would shake for 25 seconds, and the Los Angeles area would experience Mercalli intensities greater than VIII.

EARTHQUAKES / NEW MADRID 1811-1812

Most of the world's great earthquakes take place along well-defined plate boundaries, with the largest typically centered in subduction zones. Some great quakes, however, strike deep in a continental interior, far from a plate boundary.

During the winter of 1811-1812, several powerful earthquakes convulsed the central Mississippi Valley, triggering effects unique in American seismic history. Known as the New Madrid earthquakes because their epicenters were near the small town of New Madrid, Missouri, at least two of the main shocks registered an estimated moment-magnitude of 8 or higher. Damage to masonry—especially chimneys, which cracked and fell down—was extensive throughout southeastern Missouri, southern Illinois, and western Kentucky, Indiana, and Tennessee. Large areas rose and even larger sections permanently subsided, transforming land into lakes or swamps. Tennessee's Reelfoot Lake, 18 miles long, was created or greatly enlarged during the third earthquake, on February 7, 1812. Naturalist John J. Audubon, in Kentucky, recalled that "the ground rose and fell in successive furrows like the ruffled waters of a lake. The earth waved like a field of corn before a breeze." In a 100-mile-long tract from New Madrid to Marked Tree, Arkansas, liquefaction occurred, causing lowlands to flood and the ground surface to crack in yawning fissures.

As the earth jerked and twisted, plumes of sand and water spurted from the ground like geysers, rising as high as 30 feet into the air and producing sulphurous odors from decayed vegetation previously buried in topsoil. Near Blytheville, Arkansas, an area of more than 25 square miles was covered by a layer of extruded sand three feet thick.

Boats and rafts plying the Mississippi River faced towering waves that reportedly rushed upstream against the normal current, swamping some craft and washing others high upon the banks of the river. As torrents of water poured back into the Mississippi, they broke off thousands of trees and swept them into the river, creating a tangle of debris mixed with boats and canoes. Whole islands sank, while dangerous new rapids and waterfalls suddenly appeared, caused by buckling and subsidence of the river channel. Scottish naturalist John Bradbury, on a boat one hundred miles south of New Madrid, was awakened about 2 a.m. on December 16, 1811, by "a most tremendous noise…equal to the loudest thunder, but more hollow and vibrating." It seemed to him that "all nature was in a state of dissolution."

The New Madrid earthquakes are chiefly remembered not only for their violent remodeling of the Mississippi Valley terrain, but also for the astonishing distances over which they were felt. The earth shook noticeably in Boston, 1,100 miles to the northeast, in Washington, D.C., 700 miles east, and in New Orleans, 460 miles to the south. About one million square miles were forcefully set in motion.

The quakes are also distinguished by their unusual number and severity. In most major earthquakes, the initial jolt is the largest and is followed by aftershocks of gradually diminishing intensity. In these earthquakes, however, there were three main shocks, widely spaced in time, and damaging aftershocks that continued intermittently for more than a year. The first

In 1811 and 1812, the most widely felt series of quakes in American history convulsed areas of the Mississippi Valley near New Madrid, Missouri. The largest jolts, estimated at greater than magnitude 8, made land shake from Canada to New Orleans. Scientists consider future large quakes inevitable in the New Madrid Seismic Zone.

N. DAK.
MINNESOTA
S. DAK.
WISCONSIN
Mississippi
MICHIGAN
MAINE
VT. N.H.
Boston
MASS.
NEW YORK
CONN. R.I.
New York
NEBR.
IOWA
Chicago
Aurora
ILLINOIS
INDIANA
OHIO
PENNSYLVANIA
Philadelphia
N.J.
Missouri
St Louis
Saline
Mines
Cairo
Ohio
Louisville
Baltimore
Washington, D.C.
WEST VIRGINIA
MD. DEL.
KANS.
MISSOURI
Charleston
New Madrid
Fulton
KENTUCKY
Richmond
VIRGINIA
OKLA.
Blytheville
Marked Tree
Dyersburg
Memphis
TENNESSEE
NORTH CAROLINA
ARKANSAS
Mississippi
SOUTH CAROLINA
TEX.
MISSISSIPPI
ALABAMA
GEORGIA
Charleston
LOUISIANA
New Orleans
FLORIDA

0   400 mi
0   400 km

Principal disturbance area
Minor disturbance area
Recorded point of vibrations and sounds
Recorded point of vibrations only

"great" earthquake hit early in the morning of December 16, 1811; the second on January 23, 1812; and the third on February 7, 1812.

No seismographs then existed to record exact measurements of these chaotic events, but an engineer and surveyor living in Louisville, Kentucky, devised a means of evaluating their

relative size and severity. Using pendulums to measure horizontal motion and springs to detect vertical movement, Jared Brooks registered 1,874 earthquakes from December 16, 1811, to March 1812. In addition to the big three, five other earthquakes were classified as "tremendous," ten as "severe," and thirty-five as "alarming to people generally."

To residents of the Mississippi Valley, it must have seemed as if their land would never be "terra firma" again. Around New Madrid, the ground shook every day through the end of 1812 and produced seismic waves in southern Illinois for an additional two years. One recent study compared the number and magnitude of the earthquakes between December 16, 1811, and March 15, 1812, with ones occurring in southern California between 1932 and 1972. Although the Long Beach and San Fernando Valley earthquakes occurred during the study period, it took California 41 years to equal the number of quakes that the New Madrid series produced in a mere three months.

Unlike many California earthquakes, which typically rupture the surface of the earth along a discernible fault, the New Madrid quakes did not cause visible displacement of the surface. Geologists believe that the earthquakes originated in a deeply buried rift valley, a crustal block that dropped approximately 500 million years ago during an abortive rifting of North America. Because the active faults are covered by thick sediments, the most accurate way to determine their location is to find out where the current hypocenters are concentrated. The fault system—which generates an average of 200 tremors annually, most too minor to be felt—extends 150 miles southward from Cairo, Illinois, through New Madrid and Caruthersville, Missouri, to Marked Tree, Arkansas. It passes through Kentucky near Fulton and through Tennessee near Reelfoot Lake, and extends southeast to Dyersburg, Tennessee.

Since the catastrophic events of 1811-1812, the New Madrid system has produced numerous smaller but damaging quakes. Between 1838 and 1976, at least 20 shocks cracked walls and overthrew chimneys. In 1895, a quake with an estimated moment-magnitude of 6.6 generated perceptible shaking in 23 states and southeastern Canada. Besides damaging buildings near Charleston, Missouri, it created a 4-acre lake. A 1909 quake, its epicenter near Aurora, Illinois, was felt over half a million square miles.

Geologists studying this seismic zone do not expect a repeat of the 1811-1812 quakes in the near future. Earthquake statistics and probabilities indicate that large events occur roughly every 550 to 1,200 years. Although another great earthquake does not seem imminent,

EASTERN UNITED STATES

■ Eastern quakes affect extraordinarily large areas as compared with Pacific Coast earthquakes, which rarely inflict significant damage more than 60 to 100 miles from their source. The difference lies in the crustal rock that seismic waves move through. The crust in the East is somewhat cooler and less ductile, so it absorbs seismic waves less than the relatively hot crust in the West. Also, the West Coast is a geologic patchwork of discrete terranes—crustal blocks of different composition and origin from the surrounding rock. The region east of the Rocky Mountains, however, is more homogeneous—composed of relatively solid granite. Earthquakes in the eastern United States send seismic waves through a crustal block of old, brittle rock. As a result, seismic waves are unimpeded by innumerable fractures and travel great distances with little diminution in intensity.

■ On August 31, 1886, a quake of moment-magnitude 7.3 struck near Charleston, South Carolina. In the region of most intense shaking, ground failure was common, producing surface cracking and what the local residents called "sand blows," fountains of water and sand shooting high into the air. In Charleston, 2,000 brick and wood-frame buildings suffered substantial damage, and more than a hundred of them eventually had to be pulled down. Sixty people were killed. The upper floors of buildings suffered minor damage as far away as New York City, 640 miles to the northeast, and Chicago, 760 miles to the northwest.

■ An earthquake with a moment-magnitude of 5 rattled northern New York State on October 7, 1983. The quake was perceptible from southern Canada to New Jersey and was the largest one to hit the region since a 1944 event. From 1720 through the 1990s, state residents have experienced more than 350 recorded earthquakes.

■ New England has also been visited by damaging earthquakes. The first major event recorded took place in 1683. Perhaps the region's strongest historic quake was one that struck Massachusetts in 1755, when chimneys fell and church bells clanged wildly in Boston. In the same year, on the other side of the Atlantic, Europe suffered its most damaging quake when Lisbon, Portugal, was virtually destroyed. The tsunamis—seismic sea waves—that lashed the city and the fire that consumed its ruins killed thousands of people. The disaster forced European intellectuals, sons of the Enlightenment, to revise their view of nature as benign. As Voltaire pointed out, Lisbon's destruction proved that we do not live in the "best of all possible worlds." Nature, the philosopher concluded, can inflict as much random violence as humanity itself.

Lightning arcs across the sky in St. Louis, Missouri, a reminder that nature manifests chaotic power even in America's ostensibly stable heartland. Sparsely populated or non-existent during the New Madrid quakes, St. Louis, Memphis, and other cities in the Mississippi Valley risk destruction when the next great quake shakes this seismic zone.

events of lesser magnitude remain a threat. Major earthquakes of magnitude 7 or higher jolt the area every 254 to 500 years, and moderate shocks of 6 or greater occur every 70 to 90 years. Today about 12 million people live in the area most severely shaken in 1811-1812. Because of high population density and the general lack of building codes to encourage quake-resistant construction, even a moderate earthquake could cause significant damage. A recent survey reveals that St. Louis County has 140,000 unreinforced masonry structures, the kind most likely to collapse during an earthquake.

Many towns and cities within the seismic hazard zone, including parts of Memphis, Tennessee, and St. Louis, Missouri, sit on sandy sediments of the Mississippi River floodplain.

The epicenters of future quakes are likely to be as close to these population centers as the epicenter of the 1989 Loma Prieta quake was to San Francisco, when ground shaking was greatly intensified as seismic waves rolled through landfill in the Marina District.

A magnitude 6 to 7 earthquake centered on the New Madrid fault system could be devastating, throwing old frame buildings off their foundations, collapsing freeway overpasses, and reducing thousands of brick buildings to heaps of rubble. Because most structures in the region are not built to withstand even a moderate earthquake, such an event could cause far more damage than it would in some western states where design standards for buildings and bridges are more rigorous. Current studies suggest that a large quake could kill thousands of people and cost tens of billions of dollars in property losses. If private or government agencies are unable to provide sufficient emergency services, the damage to the social fabric could be overwhelming.

EARTHQUAKES / TANGSHAN

At 3:42 a.m., on July 27, 1976, a moment-magnitude 7.4 earthquake obliterated Tangshan, a densely populated industrial city in northeastern China, killing 655,000 to 800,000 people. Although not as powerful as the subduction-zone earthquakes that jolted Chile and Alaska in the 1960s, the shock was far more devastating in terms of human suffering and losses.

At least four factors contributed to the unusual destructiveness. The hypocenter was located beneath a highly developed urban area, causing strong ground movement throughout the city. Shaking intensities were magnified by the soft fluvial sediments where Tangshan was built, triggering widespread liquefaction and ground failure. In addition, a large portion of the region's inhabitants lived in multistory brick apartment buildings or single-story brick or adobe houses, up to 95 percent of which collapsed in areas undergoing the most intense shaking. Finally, because the quake occurred when most people were sleeping, hundreds of thousands were trapped and crushed when their poorly designed homes fell on them.

In seconds, a thriving industrial center was reduced to rubble. Even "quake-resistant" steel-and-concrete buildings suffered total collapse in areas that experienced maximum shaking. When a 7.1 aftershock struck later in the afternoon, the majority of structures that had withstood the main temblor also crumbled.

The horrors faced by Tangshan's people are unimaginable. A heat wave brought suffocating humidity, followed by incessant rain that flooded lavatories, sewers, and gutters. As unburied bodies putrefied and sanitation conditions worsened, fears of an epidemic afflicted a populace already traumatized by shock and grief. Outbreaks of dysentery, typhoid, influenza, and encephalitis spread among refugee camps. Although government relief was provided as soon as possible, many days passed before adequate stores of water, food, medicine, and clothing reached most residents.

The social consequences extended far beyond Tangshan. When 19 provinces, all seismically vulnerable, also issued earthquake warnings, the general population's latent fear of the earth reportedly emerged as a national phobia, creating a panic that swept across China. Millions of people left their houses and moved to shelters set up on roadsides. In Beijing, 700,000 residents camped out in tents, apparently convinced that an earthquake would soon demolish the city.

Although Tangshan was largely rebuilt within a decade after the quake, the psychological wounds were not easily healed. Ten years later, many survivors still could not break free of the terrors. One woman trapped in the rubble for three days developed overwhelming claustrophobia. Every time the sky grew dark, as it was during the earthquake, she would rush outdoors, complaining that she was suffocating.

Almost all of China is subject to damaging earthquakes, with 21 of its 30 provinces experiencing a total of 648 jolts of magnitude 6 or greater this century. Between 1949 and 1986, some 100 earthquakes killed 27.3 million people and injured 76.3 million more, a toll surpassing that exacted in any other industrialized country.

Given China's vulnerability, it is not surprising that its scientists have pioneered the art

After the earthquake of July 27, 1976, the city of Tangshan, China, resembles ground-zero of a nuclear attack (left, upper). The quake killed an estimated 655,000 people and reduced most buildings in the area to heaps of rubble. A year earlier, Chinese scientists successfully predicted an earthquake in Haicheng, but they failed to detect conclusive precursors in this case. Ten years after its destruction, Tangshan appears largely rebuilt, with quake-resistant high-rises and open squares (left, lower).

CHINASTOCK (BOTH)

Although many buildings were damaged, few people died.

Marking the first time that scientists had predicted a major quake, the forecast seemed to be a turning point. Based on ground deformation, changes in groundwater level, tiny foreshocks, and animal behavior, the prediction indicated that scientists had found a key to reducing the destruction of earthquakes.

Unfortunately, the precursors signaling the Haicheng event did not show up before the Tangshan catastrophe. Chastened, scientists found that different seismic faults, like individual volcanoes, can exhibit a wide range of behaviors. Sometimes quakes occurred without preliminary anomalies; sometimes apparent precursors took place with no earthquake following. Although scientists may someday be able to detect a pattern of precursory signals characteristic of particular faults, enabling them to issue warnings that could save innumerable lives, that ability now seems to lie in the distant future.

of earthquake prediction. After the Xingtai and Ningjin earthquakes of March 1966, in which more than 8,000 people died, Premier Zhou Enlai challenged geologists to find a way to forecast large events. Nine years later, seismologists predicted a sizable earthquake would occur near Haicheng in early 1975. On February 4, they telephoned government agencies that a quake was imminent, resulting in immediate evacuation. At 7:36 that evening, just as people were about to return home to escape the winter chill, a magnitude 7.3 shock convulsed the area.

EARTHQUAKES / KOBE

Japan has experienced numerous destructive earthquakes throughout its long recorded history. As a result, engineers here are keenly aware of seismic dangers and take great pains to design some of the world's most earthquake-resistant structures, including high-rise office buildings and elevated roadways. After studying the effects of the 1994 Northridge earthquake near Los Angeles, California, many Japanese scientists were apparently reassured that their highways and bridges were built to withstand a similar jolt.

On the first anniversary of the Northridge quake, a larger one struck near Kobe—Japan's second most important port—and painfully demonstrated how vulnerable even a modern industrialized city is to a near-direct seismic hit. At 5:46 a.m. on January 17, 1995, a moment-magnitude 6.9 quake caused appalling damage, leaving approximately 5,200 persons dead, more than 190,000 buildings destroyed or damaged, and 300,000 people homeless. Damages totalled 200 billion dollars.

Although the quake epicenter was located about 15 miles southwest of downtown Kobe, the fault movement may have extended beneath the city itself. Built on a narrow strip of land along Osaka Bay, Kobe sits atop loose, water-logged bay sediment, the kind of soil that amplifies the effects of seismic waves as they pass through. Exceptionally strong shaking triggered widespread liquefaction and ground subsidence all along the margins of Osaka Bay.

Elevated roads and railways failed catastrophically; their supports collapsed for want of steel reinforcement. A 1,650-foot-long section of the elevated Hanshin Expressway, which is the main vehicular traffic artery through Kobe, rolled over on its side, tumbling cars, buses, and trucks to the ground. Two other sections of the expressway also collapsed, as did numerous overpasses and railroad bridges, leaving railway lines dangling in midair. All of the major transportation systems connecting western Japan with the rest of the country were severed.

Most of the deaths and serious injuries took place when older wood-frame and stucco houses with clay-tile roofs collapsed. Some modern, concrete-and-steel, industrial edifices suffered heavy damage, including multistory buildings that fell on their sides like play houses kicked over by an ill-tempered giant. The vast majority of buildings dating from the 1980s fared reasonably well. The ones that collapsed did so because the ground apparently gave way beneath them. One six-story bank building leaned into the street like a modern tower of Pisa. Sidewalks and street pavements cracked and sank several feet, victims of extensive ground deformation. As in other earthquakes affecting areas with high groundwater levels, the violent shaking ejected fountains of water and soil, leaving open areas pockmarked by miniature craters and blanketed with wet sand.

Widespread ground and building failure ensured that Kobe's underground utilities were wrecked, disrupting services that provided gas, water, and the disposal of sewage. For days after the quake, the majority of households had neither water nor gas, and many residents had no electricity. The lack of water was not only a hardship for the population as a whole, but also contributed to the rapid spread of fires that broke out along Kobe's ravaged waterfront and

As if uprooted from its foundation and flung down by a bellicose giant, a modern building lies on its face in downtown Kobe, a victim of the widespread ground failure generated by 20 seconds of intense vibrations during the 1995 quake. Seismic waves cause water-logged soils to liquefy, robbing them of their ability to support heavy structures, a hazard faced by almost every port built on landfill along the Pacific Rim.

in the downtown area. At least a dozen major fires raged out of control for up to 48 hours, burning whole neighborhoods to the ground.

Some of Kobe's citizens, many of whom had no shelter in near-freezing temperatures, were openly critical of what they considered government indifference to their plight. When questioned by reporters, they angrily denounced civil authorities for their failure to provide adequate relief, including basics such as water, food, blankets, tents, or medical supplies. Others attacked elected officials' reluctance or refusal to accept foreign aid. Although rescue workers from Switzerland were allowed to bring dogs trained to find bodies buried in the rubble, many other nations' offers of assistance were summarily rejected.

Lessons learned from the Kobe earthquake are sobering. Although Japanese construction standards are perhaps the best in the world, damage was astonishingly extensive. With so many buildings destroyed and transportation corridors severed, some engineers reportedly

lost confidence in the human ability to create bridges, highways, or other edifices that could ride out a moment-magnitude 7.9 quake, such as the one that devastated Tokyo in 1923.

Perhaps the most significant lesson involves not only how well buildings are constructed but also where. Codes that permit large commercial or industrial developments to be built on landfill or other soft ground may have to be drastically revised. When shaking causes soil to become no more solid than mush, even carefully designed edifices can sink, tilt dangerously, or collapse altogether. Populous cities all along the Pacific Rim, from Kobe and Yokohama, in Japan, to San Francisco and Oakland, in California, stand largely or in part on bay mud that will rapidly liquefy during the next major earthquake. Because it is probably not feasible to expect governments to move these or other busy ports from their present locations on unstable ground, disasters similar to the one suffered by Kobe seem to be an inevitable part of these cities' seismic futures.

TSUNAMIS

Although they are often called tidal waves, the great surges of water that in 1964 ravaged the coastlines of Alaska, Oregon, and northern California had nothing to do with tides, which are regulated by the gravitational attraction of the sun and moon. Properly known as a seismic sea wave or tsunami—the Japanese term meaning "great harbor wave"—this phenomenon is caused by sudden uplift or subsidence of the seafloor, massive underwater landslides, or volcanic eruptions. When a quake, for example, raises or drops the seafloor, it displaces water in the affected area, setting in motion waves that speed across the ocean at 500 to 600 miles an hour.

Tsunamis crossing the open ocean are usually only a foot to six feet high and are virtually undetectable at sea. When one approaches land, however, its bottom slows as it drags along the increasingly shallow ocean floor. By contrast, the wave's top continues at full speed and, when the seafloor topography is favorable, it can rise 30, 50, or 100 feet. The highest tsunami on record crested at 278 feet as it swept through the Ryukyu Islands south of Japan in 1971.

Because lands near the Pacific Ocean experience most of the world's great earthquakes and explosive volcanic eruptions, the tsunami threat is particularly acute along the coasts of Alaska, Russia, Japan, the Philippines, Indonesia, and western South and Central America. In July 1993 an earthquake in the Sea of Japan produced tsunamis that inundated Japan's northwestern coast, killing more than 120 people. On April 1, 1946, a Richter magnitude 7.2 earthquake south of Alaska's Unimak Island sent waves 20 to 32 feet high into Hilo,

Hawaii, 2,500 miles from the epicenter. The tsunamis obliterated Hilo's waterfront, flooded the business district, and drowned 159 persons.

Some of the most destructive waves are generated by volcanoes. In August 1883, Krakatau, a volcanic island in Indonesia's Sunda Strait, exploded unexpectedly, sending a cloud of tephra and aerosols more than 30 miles into the stratosphere. As the summit shattered and collapsed, tsunamis swept the coasts of Java and Sumatra, destroying 165 settlements and drowning 36,000 people.

In the prehistoric past, even larger tsunamis formed when huge flanks of the Hawaiian volcanoes collapsed, creating landslides that spread up to 125 miles across the ocean floor. Waves cresting a thousand feet scoured shorelines, eroding steep cliffs hundreds of feet high. Today, part of Kilauea's south flank, riddled by

KIMIMASA MAYAMA/REUTERS/CORBIS-BETTMANN

Seismic sea waves ravage the Pacific Rim. In July 1993 an earthquake triggered a tsunami and fires, almost obliterating the town of Aonae on the Japanese island of Okushiri (above). Carried ashore by a tsunami that hit the Indonesian island of Flores in December 1992, dead tuna lie amid the wreckage of a waterfront street (opposite). Virtually every nation bordering the Pacific faces episodes of comparable destruction.

extensive fracturing, slumps seaward periodically. When it collapses and avalanches onto the ocean floor, it will produce huge tsunamis.

After the 1946 tsunami swept into Hilo, scientists developed the Tsunami Warning System, which has its headquarters in Honolulu. The Pacific Tsunami Warning Center is connected to a network of seismographs and tide-gauge stations around the ocean. Within 30 to 40 minutes after an earthquake with the potential to create a tsunami occurs, the center issues a "tsunami watch," giving an estimated time of arrival. The

first evidence for a tsunami's existence comes from tide stations near a quake's epicenter. Once a wave is confirmed, the Honolulu office puts out a "tsunami warning." When an official warning is issued, local authorities can evacuate low-lying coastal areas. Because a tsunami's velocity is directly related to water depth, its arrival time from any point on the Pacific Rim can be predicted. Its size, however, is difficult to forecast precisely.

The behavior of a tsunami and its effects on different coastlines vary greatly. The size and destructiveness of the wave depend on local topography and the direction from which the wave approaches. Whereas a tsunami caused by an earthquake in the Gulf of Alaska may produce only ripples in Japan, it may generate waves cresting at 50 feet or higher along the coast of California.

TSUNAMIS / ALASKA 1964

As if heaving earth, collapsing buildings, and raging fires were not sufficiently destructive effects of the great earthquake that struck Alaska in 1964, nature wielded yet another lethal weapon—seismic sea waves. Consider the triple blow of earthquake, fire, and flood that devastated Seward, a small port and railroad terminus on the Kenai Peninsula. When a moment-magnitude 9.2 shock, the strongest ever recorded in North America, struck at 5:36 p.m. on March 27, vast areas of the ocean floor were abruptly raised many feet, while others tipped and sank, setting in motion enormous quantities of seawater.

A 30-foot-high wall of water slammed into the waterfront at Seward. Gene Kirkpatrick watched in horror as fire and water united to create a holocaust. "The lid blew off in the [oil] storage tank area almost the first thing," he recalled. "Then, when the fire was really roaring, the wave came up Resurrection Bay there and spread it everywhere. It was an eerie thing to see—a huge tide of fire washing ashore, setting a high-water mark in flame, and then sucking back."

Slick with burning oil, the sea wave swept inland at one hundred miles an hour, racing over railroad tracks and into the port's east end. Scooping up locomotives, boxcars, oil drums, pier pilings, and other debris, the fiery flood rushed past the waterfront, dock area, and railroad yard far into town. A series of waves rolled through in rapid succession and, as they withdrew, carried most of what had been Seward out to sea. In the cold darkness, Jim Kirkpatrick, Gene's stevedore brother, could see flaming wreckage dotting Resurrection Bay like so many funeral pyres. "A lot of the snapped-off pilings were floating around upright, because their lower sections were waterlogged. Then the top sections with all that coating of tar and oil caught fire," igniting mournful beacons.

Waves swept coastal settlements along the Gulf of Alaska, extending their lethal effects as far south as Crescent City, California, where a dozen people drowned when water 9 to 20 feet high rolled inland. The Alaskan port of Valdez, which lies at the head of a narrow inlet about 30 miles from Prince William Sound, suffered a high loss of life. When the quake struck, about 30 stevedores and onlookers were standing on a pier; within seconds, people and pier had vanished. As a 30-foot wave approached, Jim Aubert saw "maybe a dozen people turn and break for the beach, but she was a long pier—maybe a hundred yards—and they hadn't the littlest chance. She was sucked under all at once...."

On Kodiak Island's east coast, the Aleut fishing village of Kaguyak was scoured from the land. First came a smooth, high swell of water, followed by the sea's withdrawal far down the shore, stranding sea creatures on a suddenly exposed beach. After the water had advanced and receded a second time, the third and largest wave rushed in, smashing every house and church.

Alerted by the initial wave, Kaguyak's two-score residents had fled to high ground, where they watched the second flood withdraw beyond the normal low-tide mark. Knowing that they would probably have to spend the night on their windy hilltop, several men dashed back into town to gather blankets, portable radios, and flashlights.

Tsunamis following Alaska's March 27, 1964, earthquake—the largest in North America in historic time—heavily damaged many coastal settlements, including Seward, a port and railroad terminus (below). Tracks of the Alaska Railroad (left), severed by land washouts and sea waves at Seward, show the power of the killer quake and tsunamis.

While the men were foraging for supplies the terrible third wave arrived. Miraculously, three of the men who saw the threat coming climbed aboard two dories beached beside the houses and managed to survive the inundation that washed the settlement into oblivion. Roger Williams, a young village leader, remarked that the Aleuts should have remembered their forebears' warning. Long ago their previous village had been similarly engulfed in a disaster that gave rise to the adage: "The third wave is the worst—watch out for the third one."

VOLCANOES

Long before modern scientists discovered how and why volcanoes erupt, ancient storytellers spun myths about mountains that thundered, belching smoke and flame. In the Mediterranean region, which contains about 15 historically active volcanoes, early classical poets speculated that eruptions were caused by supernatural beings who inhabited volcanic landforms. Some Greek writers accounted for Mount Etna's frequent rumblings and lava flows by ascribing them to the noisy struggles of Typhoeus, the fiery dragon Zeus had imprisoned under Etna's bulk. According to one tradition, when Hephaestus, the god of fire and metalcraft, was temporarily exiled from heaven, he set up his forge on the Aegean island of Lemnos, boisterously pumping his bellows and lobbing hot rocks into the air. The Romans insisted that their counterpart of Hephaestus, Vulcan, established his overheated workshop under an island near the coast of Sicily—Vulcano—the name from which our word "volcano" is derived.

For the pre-Columbian Indians of northern California and southern Oregon, the towering glacier-shrouded cone of Mount Shasta was the earthly tepee of Skell, the benevolent sky god who sporadically visited mortals, announcing his presence by kindling a spark-shooting blaze at Shasta's summit. Skell's chief rival, an underworld deity named Llao, occupied another peak in the Cascade Range 125 miles to the north. During a climactic battle between Llao and Skell, Llao's own mountain fell on top of him, permanently confining the ill-tempered deity underground. The huge caldera that formed when Llao's former peak collapsed, a circular depression 5 by 6 miles in diameter and 4,000 feet deep, was gradually filled by snowmelt and rain, creating the breathtaking indigo of Oregon's Crater Lake.

Although we no longer believe that divine beings set up shop inside active volcanoes, we are likely to find that experiencing a great volcanic eruption is just as frightening and awe-inspiring as it was in the days of our remote ancestors. The Indian tribes who witnessed the cataclysmic outburst that overnight transformed a lofty cone, the 12,000-foot-high Mount Mazama, into the deep basin now holding Crater Lake were so impressed by the event that they preserved traditions describing it for scores of generations.

Some catastrophic eruptions that took place in this century have made similarly indelible impressions on the modern consciousness, including that of Mount Pelée, which destroyed the Caribbean city of St. Pierre on May 8, 1902. Mount Pelée's eruption is memorable not only for its tragically high death toll—30,000 persons—but also because Pelée demonstrated that volcanoes can erupt in extremely destructive ways that scientists had not previously anticipated.

In 1902, most geologists believed that volcanoes generally behaved in one of two ways. Explosive eruptions, such as those of Vesuvius or Krakatau, shot a column of ash-laden steam high into the air and showered the countryside with falling debris. In effusive activity, such as that in Hawaii, volcanoes more or less quietly emitted streams of lava that typically flowed serenely downslope and into the sea.

Mount Pelée's eruption in 1902, however,

Geologists sample gases issuing from sulfurous fumaroles on Vulcano, a volcanic island off the west coast of Italy. Ancient Romans said their fire god, Vulcan, kept his workshop deep within this mountain. Sudden increases in heat or gas emission may warn scientists of magma rising toward the earth's surface.

was quite different. Instead of only rising vertically, the eruption column, composed of superheated gases and incandescent rock particles, hugged the ground, racing horizontally down Mount Pelée's flanks and through St. Pierre, flattening thick stone walls, incinerating wooden buildings, and asphyxiating all but two of its inhabitants, including a convicted felon whose confinement in an underground jail cell probably spared his life.

Only a few people aboard ships anchored in St. Pierre's harbor survived to describe the holocaust, including assistant purser Thompson, an officer on the steamship *Roraima*. From his ship's deck, Thompson watched as Pelée suddenly transformed the port into a raging inferno.

"There was a constant muffled roar. It was like the biggest oil refinery in the world burning up on the mountaintop. There was a tremendous explosion about 7:45 [a.m.].... The mountain was blown to pieces. There was no warning. The side of the volcano was ripped out, and there hurled straight towards us a solid wall of flame. It sounded like a thousand cannon. The wave of fire was on us and over us like a lightning flash. It was like a hurricane of fire. I saw it strike the cable steamship *Grappler* broadside on and capsize her. From end to end she burst into flames and then she sank. The fire rolled in mass straight down on St. Pierre and the shipping. The town vanished before our eyes and then the air grew stifling hot and we were in the thick of it.

"Wherever the mass of fire struck the sea, the water boiled and sent up great clouds of steam.... The fire swept off the [*Roraima's*] masts and smokestack as if they had been cut by a knife.... Before the volcano burst, the landings of St. Pierre were crowded with people. After the explosion, not one living being was seen on the land."

Although it seemed to the assistant purser that the blast tore Mount Pelée apart, what he actually saw was a laterally directed cloud of gas and incandescent ash rushing down Pelée's slopes straight toward him. When the French volcanologist Alfred Lacroix later investigated the eruption, he christened Pelée's deadly cloud a *nuée ardente*—a "glowing cloud." Also called a pyroclastic flow, the nuée ardente that annihilated St. Pierre's population is now recognized as a common volcanic phenomenon, albeit one of the most dangerous.

How Big Was the Blast?

Although volcanologists do not have instruments to measure the relative magnitude of volcanic eruptions, they can calculate the relative strength of individual eruptions by considering such factors as the volume of material ejected, the distance to which large rock fragments are hurled, and the height of the eruptive cloud. To compute the comparative size of prehistoric outbursts, for which there are no written records, and that of eruptive events reported by eyewitnesses, scientists have devised the Volcanic Explosivity Index (VEI). Like the Richter scale for measuring earthquakes, the VEI has a simple numerical rating system, ranging from zero to eight, with each successive full number representing an increase of about a factor of ten. The higher the number, the more voluminous and violently explosive the eruption.

In a recent edition of the Smithsonian Institution's *Volcanoes of the World,* a comprehensive listing of the world's 1,511 known volcanoes, the editors assign no eruptive event during the last 10,000 years a VEI rating of 8.

Only four, including the Tambora eruption of 1815, receive a VEI rating of 7. But even Tambora, the Indonesian volcano that produced the most voluminous outburst of modern times—about 20 cubic miles of dense magma—is easily overshadowed by another Indonesian volcano, Toba, which ejected more than 600 cubic miles of magma approximately 74,000 years ago, forming enormous pyroclastic flows and surges that smothered at least 10,000 square miles in deposits up to 1,000 feet thick. Toba's ash clouds may have created near-global darkness and caused a severe worldwide cooling, perhaps, as some geologists speculate, precipitating a new pulse of the last ice age.

North America's Yellowstone volcano produced a comparable eruption about two million years ago. Fortunately for humanity, no eruption of this size has occurred in historic time. During recent decades, the world's volcanoes have produced about 60 eruptions per year. Of these, one eruption every few weeks rates a VEI rating of 2, meaning that it generates about 1.3 million cubic yards of rock fragments. A VEI-3 eruption producing about 13 million cubic yards of tephra, such as the 1985 Nevado del Ruiz event that killed 23,000 people, occurs several times a year.

Only once in a decade do VEI-5 events such as the 1980 eruption of Mount St. Helens, which ejected about .25 cubic mile of magma, take place. An outburst like that of Krakatau in 1883, which had a volume of 2.4 cubic miles—10 times that of St. Helens—rates a VEI 6, and occurs about once in a hundred years. As the Smithsonian editors point out, St. Helens devastated hundreds of square miles, creating a

Atop Mount Unzen, one of Japan's deadliest volcanoes, a lurid glow outlines the profile of a growing lava dome. In June 1991, incandescent rock avalanching from a sudden collapse of the dome generated a pyroclastic flow and surge that snuffed out the lives of 41 people, including volcanologists Katia and Maurice Krafft and Harry Glicken. An eruption in 1792 killed approximately 15,000 persons.

local catastrophe, whereas Krakatau was a regional catastrophe, killing 36,000 people, one as far away as India.

Tambora's explosion in 1815 had global consequences, significantly lowering temperatures worldwide and causing the "year without a summer," when nightly frosts blighted crops from Europe to New England. We can only imagine the repercussions that a repetition of the Yellowstone explosion of two million years ago would have today. The event distributed more than 600 cubic miles of ash over a 16 state area, including the future sites of Los Angeles, Tucson, El Paso, and Des Moines. It is not reassuring to know that such gigantic pyroclastic flows, which have sporadically punctuated the earth's long history, commonly follow quiet intervals hundreds of thousands of years long.

VOLCANOES / SHIELDS

From tiny heaps of cinder to towering cones whose snowy summits pierce the clouds, volcanoes vary tremendously in size, shape, and eruptive behavior. A volcano's form, whether steep and symmetrical like Japan's Fujiyama or broad and gently sloping like Hawaii's Kilauea, is generally determined by the kind of eruption—explosive or effusive—it produces. The volcano's eruptive style, in turn, is largely dependent on the chemical composition of the lava it erupts.

The chemical and mineralogical constituents of nearly all kinds of lava are virtually identical—silica and oxides of calcium, sodium, aluminum, potassium, sulfur, magnesium, and iron—but they vary considerably in their proportion. The silica content mainly determines the group to which a type of lava belongs. Basaltic lavas, which form the ocean floors and mid-oceanic volcanoes, are low in silica (about 50 percent by weight) and high in iron and magnesium, which gives them their dark color. When such lavas are erupted at high temperatures, they can be extremely fluid, spreading out in thin sheets and traveling great distances from their source.

Volcanic mountains built almost exclusively by thin lava flows, such as the ones in Hawaii and some in Iceland, are called shield volcanoes. Named for the supposed resemblance to a Roman soldier's shield laid flat with the curved side up, shield volcanoes have low, undramatic profiles. Composed of thousands of basalt flows, Hawaii's Mauna Loa is the world's largest mountain, with a volume of nearly 10,000 cubic miles. Along with its neighbor, Mauna Kea, it is also one of the earth's two

FRED HIRSCHMANN

tallest mountains, rising about 32,000 feet from its base on the ocean floor.

The Hawaiian style of eruption, characterized by jetting fountains of basaltic lava and incandescent rivers of molten rock, is considered the least dangerous type of eruptive activity. Instead of fleeing when Kilauea erupts, tourists flock to watch its pyrotechnic displays, giving it the reputation of a "drive-in" volcano. Kilauea's admirers may feel confident because almost any able-bodied person can outrun the average lava flow. Except near its source or on a very steep slope, where lava streams can move at speeds up to 20 miles an hour, the flow rarely advances faster than about one mile an hour.

Although they seldom endanger people, Hawaiian lava flows commonly destroy property, burying fields of sugarcane and seaside villages under acres of black rubble. In 1935 and 1984, voluminous flows pouring from Mauna Loa threatened to engulf Hilo, the largest city

A brilliant lava fountain (left), propelled by gas escaping from fluid basaltic magma, sprays from a cinder cone on the flanks of Kilauea, one of the five overlapping shield volcanoes that form the Big Island of Hawaii. In almost continuous eruption since January 1983, Kilauea has buried several residences and segments of the Chain of Craters Road under acres of new lava. The surface of a solidified flow (opposite) resembles a petrified mass of coiled rope.

PAUL CHESLEY

on the island of Hawaii. Although the port city was spared when eruptions abruptly ceased, it is probably only a matter of time before Hilo is paved over. Since January 1983, Kilauea has been in almost continuous eruption, the longest in recorded history, emitting floods of lava from its east rift zone, a system of deep fractures in Kilauea's eastern flank.

Geologists use Hawaiian terms to classify two common kinds of lava flow: 'a'ā and pāhoehoe. Always basaltic in composition, pāhoehoe displays a smooth, undulating surface that resembles coiled rope. In a pāhoehoe flow, lava creeps over level ground at the rate of about three feet a minute, advancing by exuding lobes of molten material. As the crust cools and hardens, forming a solid rocky canopy insulating the molten interior, liquid lava travels through pāhoehoe in a series of tubes. When the lava drains away from the interior at the end of an eruption, it leaves long hollow tunnels inside

the solidifying lava. These tunnels are exposed only when a section of the overlying crust collapses, opening an entrance to the tube within.

Whereas pāhoehoe flows can travel many miles from their source, 'a'ā flows typically move sluggishly and travel shorter distances. 'A'ā is characterized by a rough, jagged crust resembling black slag from a smelter. About 99 percent of the Big Island of Hawaii is made of 'a'ā and pāhoehoe; one percent consists of tephra (airborne fragments) from explosive eruptions.

Hawaiian volcanoes can occasionally stage temper tantrums, as Kilauea proved in 1924, when groundwater seeped into its lava-filled crater, triggering a series of steam blasts that killed a photographer. Much larger explosive outbursts, generating pyroclastic flows and surges, have occurred sporadically in the prehistoric past, suggesting that volcano watchers cannot invariably count on Kilauea and Mauna Loa to keep their fireworks under control.

VOLCANOES / CINDER CONES

Mildly explosive eruptions are called Strombolian, after Stromboli, an island volcano in the Tyrrhenian Sea off the coast of Sicily. Known since classical times as the "lighthouse of the Mediterranean," Stromboli erupts with almost clockwork regularity, guiding mariners with a "pillar of cloud" by day and a "pillar of fire" by night.

An almost ubiquitous volcanic landform, the cinder cone is built by Strombolian activity. Found everywhere from central California and the deserts of Arizona to the lava wastes of Iceland, cinder cones are formed when moderately explosive eruptions spew chunks of partly molten rock into the air. Known as tephra, these airborne fragments—ranging in size from large blocks many feet in diameter to ash particles less than a tenth of an inch across—fall back around the erupting vent, piling up to form a steep-sided cone.

In this century, geologists have witnessed the birth of several cinder cones. The most famous was Parícutin, which unexpectedly sprouted on a farm in Mexico on February 20, 1943. Following a series of earthquakes, a hole yawned in Dionisio Pulido's cornfield and began belching steam and hot cinders. Within 24 hours, an andesite-tephra cone about 33 feet high occupied the field. Named for the village of Parícutin, which it later buried under a lava flow, the fire-mountain grew rapidly. After the first week it stood 550 feet tall; after one year, it reached more than 1,100 feet.

Over the next several years, Parícutin grew mainly in volume. As loosely compacted piles of volcanic rubble, cinder cones are unstable. Because of the cone's structural weakness, a heavy column of molten rock cannot rise high inside the edifice without causing it to collapse. Most of Parícutin's lava flows thus emerged from vents near the foot of the cone, some spreading several miles from their source. In June 1944, a thick stream bulldozed the village of San Juan Parangaricutrio, crushing its houses and partly burying its church, leaving only the church bell tower and a small part of the facade rising above a chaotic waste of lava.

After nine years of almost continuous activity, Parícutin abruptly ceased erupting in 1952. This short life-cycle characterizes cinder cones, which typically form in a single eruptive episode and then fall silent. If fresh magma rises to the surface in the same vicinity, it does not reactivate existing cones, but blasts open a new vent and creates an entirely new volcano.

The hazards of living near a geologically young cinder cone were vividly illustrated on January 23, 1973, when the residents of Vestmannaeyjar, on the island of Heimaey off the south coast of Iceland, were jolted awake by the bellowing of a newborn volcano. A fissure had opened less than a mile outside of town, near the foot of Helgafell, a large cinder cone that had formed about 6,000 years earlier. Acting with composure and efficiency, within 6 hours officials of Vestmannaeyjar transported more than 5,000 residents to the mainland, leaving only 300 volunteers to defend the town against fires ignited by showers of red-hot bombs.

Helgafell's noisy sibling, which soon grew to a height of 735 feet, ejected basaltic tephra that quickly shrouded the island. In the neighborhood nearest the volcano, 300 houses went up in smoke while another 65 were entombed

In January 1973 the citizens of Vestmannaeyjar, on Iceland's Heimaey Island, awoke to find a new volcano blazing just outside town. For the next several months, the inhabitants battled the growing cinder cone, digging their houses out of drifts of black ash and spraying water on lava that threatened to block their harbor.

in charcoal-gray ash deposits up to 20 feet thick. As the eruption continued, lava streams threatened to close the harbor's only entrance, an eventuality that would prove fatal to the town's reason for existence, its fishing industry.

Aided by Icelandic scientists, for the next four months the islanders battled the lava by spraying it with seawater. An elaborate system of pumps, pipes, hoses, and sprinklers was set up in the hope that by rapidly chilling the lava, the movement of the flow could be slowed. When the lava had come within 150 yards of sealing the harbor entrance, the eruption dwindled and stopped. Drenching the flow with seawater may have helped solidify its margins enough to retard movement, but whether it influenced the flow's ultimate size is not certain.

Although the people of Vestmannaeyjar had to dig their town out of 1.3 million tons of gritty ash to reoccupy their homes, the new cinder cone also provided some benefits. Besides adding almost one square mile to their island, its lava constructed a new breakwater, making the harbor more secure than before.

VOLCANOES / COMPOSITES

Composite volcanoes produce some of the world's most violent eruptions. The 1883 explosion of Krakatau (below), an island volcano in Indonesia, killed 36,000 people. A cone on the Kamchatka Peninsula belches tephra (opposite).

PAUL CHESLEY

The composite cone, or stratovolcano, is a volcanic structure built of effusive lava flows and explosively ejected tephra. Some of the loftiest and most symmetrical peaks, including Fujiyama in Japan, Mount Mayon in the Philippines, and Mount Shishaldin in Alaska, are composite cones. Although less symmetrical, Washington State's Mount Rainier, California's Mount Shasta, Italy's Mount Vesuvius, and Mexico's Popocatépetl are also of the composite variety. Because they typically alternate between quietly emitting lava and exploding violently,

composite cones are among the most unpredictable and dangerous of all volcanoes.

Among the deadliest phenomena associated with composite volcanoes is the pyroclastic surge. Commemorating the unspeakable carnage wrought at St. Pierre, geologists call an eruption that produces a laterally directed surge Peléan, after its prototype at Mount Pelée. A Peléan eruption occurs when thick, sticky masses of lava ooze from a vent located high on a steep volcanic cone. Too viscous to flow far, the lava piles up to form a mound over the vent called a lava dome. As the lava dome grows, it

188

becomes unstable. Huge sections of the dome can suddenly break away to form an avalanche of lava fragments that are mostly solidified but still partly molten. As the fragments rapidly disintegrate, a pyroclastic flow is formed, sending searing-hot gases and lava fragments racing downslope with hurricane force.

On June 3, 1991, a dome atop Japan's Mount Unzen exploded, creating a surge that raced down the volcano's flanks and killed 41 people. Among them were French volcanologists Katia and Maurice Krafft and U.S. geologist Harry Glicken, who had come to study the eruption.

The tragic deaths of these scientists are rife with irony. The Kraffts, who for years had traveled around the globe photographing volcanic eruptions, had just finished preparing a videotape, *Understanding Volcanic Hazards,* which was widely distributed among U.S. military leaders and geologists working at Clark Air Base in the Philippines. According to members of the U.S. Geological Survey, the Kraffts's film was instrumental in alerting authorities to the dangers posed by Mount Pinatubo, which was then in the preliminary stages of what turned out to be a catastrophic eruption. Influenced by the Kraffts's work, at the time of the filmmakers' deaths Philippine and U.S. agencies were laying plans to evacuate populations living in Pinatubo's shadow, arrangements that were to save many thousands of lives.

The story of Harry Glicken is even more poignant. As a graduate student in the spring of 1980, he worked for the U.S. Geological Survey team at Mount St. Helens, monitoring the recently reawakened volcano while camped atop Coldwater Ridge about six miles northwest of St. Helens's summit. On May 17, David A. Johnston, also of the USGS, volunteered to change places with his friend so Glicken could take a break from his solitary duty.

At 8:32 the next morning, May 18, Mount St. Helens erupted violently, ejecting an immense pyroclastic surge that swept over 230 square miles north of the volcano, killing Johnston and 56 others. Although reportedly burdened with survivor's guilt, Harry Glicken continued his career in volcanology, only to perish 11 years later in the same kind of deadly eruption that had taken his colleague.

VOLCANOES / VESUVIUS A.D. 79

From his family's villa at Misenum on the glittering Bay of Naples, an 18-year-old youth gazed across "the loveliest regions of the Earth." Here, on soils enriched by volcanic ash, were prosperous farms, lush vineyards, and luxurious summer homes of wealthy Romans who had escaped the muggy capital to enjoy coastal sea breezes. On this early afternoon of August 24, A.D. 79, Pliny the Younger found only one feature to mar the landscape.

Pliny's mother called her son's attention to a rather ominous sight, "a cloud which had appeared of a very unusual size and shape." As Pliny later wrote to Tacitus, the major historian of imperial Rome, the cloud took the form of a pine tree common to the area, "for it shot up to a great height" like a "very tall trunk, which spread itself out at the top into a sort of branches." From their location some 20 miles away, it was not immediately clear from which mountain the cloud arose, but it was soon determined to be Vesuvius. After sleeping peacefully for centuries, the volcano had awakened.

Young Pliny's uncle and guardian, Pliny the Elder, who was then stationed in Misenum as commander of the Roman fleet, was also a celebrated natural historian. His curiosity piqued, the older man determined to investigate the phenomenon, a resolve strengthened when he received an urgent message from a friend, the lady Rectina, whose villa stood at the foot of Vesuvius. Inviting his nephew to join him—an offer the youth politely declined, preferring to continue his studies—Pliny the Elder set out to rescue not only Rectina but also people living in various towns along the bay's coast. Ordering the galleys launched, the commander sailed toward the place from which "others fled with the utmost terror."

In his two letters to Tacitus, the oldest surviving descriptions of a volcanic eruption, Pliny apparently relied partly on the accounts of survivors who accompanied his uncle on this mission of mercy. As Commander Pliny's ships approached Vesuvius, they not only encountered a hail of hot lava fragments, "black pieces of burning rock," but also found it impossible to dock. Earthquakes had raised the shore level, and the sea had retreated almost half a mile. Accordingly, Pliny gave orders to steer toward Stabiae [now Castellammare], where he spent his final night at the house of a friend.

Uncertain whether to remain indoors and risk being crushed by falling walls (powerful earthquakes rocked the earth unceasingly) or to take his chances outside under a rain of blocks and cinders, Pliny and his companions tied pillows on their heads with napkins and braved the whirlwind of hot ash that descended on them. Although the sun had risen, "a deeper darkness prevailed than in the thickest night." Exhausted by the ordeal, the stout, 56-year-old Pliny lay down "upon a sail cloth, which was spread for him," drank some water, and perished—"suffocated," his nephew supposed, "by some gross and noxious vapor." When the eruption abated on the third day, Pliny's body was found undisturbed, "looking more like a man asleep than dead."

Perhaps because Tacitus was interested primarily in the circumstances of Pliny the Elder's death, the younger Pliny's letters, written several years after the event, do not mention the fate of the two cities buried in the eruption,

Shadowed by Mount Vesuvius, the modern town of Ercolano sprawls over the site of Herculaneum. Archaeologists have freed much of the ancient city from the pyroclastic deposits that buried it during the cataclysm of A.D. 79. Quiet since its last eruption in March 1944, Vesuvius could pose a major threat to the congested Bay of Naples region, now home to millions of people.

Pompeii and Herculaneum. Eyewitness details that Pliny provides, however, are remarkably accurate and have helped modern investigators reconstruct the sequence of eruptive events that transformed one of the most affluent and verdant areas in the Roman Empire to a moonscape of gray ash.

By identifying the series of distinctive ash and pumice deposits that Vesuvius spread over Pompeii and Herculaneum from August 24 to 25 and correlating them with particular phenomena that Pliny mentions, volcanologist Haraldur Sigurdsson of the University of Rhode Island was able to determine that the eruption progressed in discrete stages. In its opening volley at about 1 p.m., August 24, Vesuvius ejected the column of steam and ash that Pliny said resembled the shape of a pine tree. For the next 11 hours, this roiling cloud rose 12 miles into the stratosphere, carrying pieces of old rock torn from the volcano's conduit and millions of tons of fresh, glassy pumice.

Prevailing winds carried the ash cloud toward Pompeii, a prosperous city of some 20,000 inhabitants, where showers of lapilli, the Latin term for "small stones," began to fall about 1:30 p.m. As the sun-extinguishing canopy extended over the city, pumice and ash rained down on Pompeii, accumulating at the rate of six inches an hour. Whereas some Pompeiians fled in terror, abandoning their household goods, others took refuge in their houses and shops, hoping that the tempest would soon blow over. Lying 4 miles upwind from Vesuvius, Herculaneum, a residential town of about 5,000 people, received only a light sprinkling of ash.

The situation changed abruptly at about 11:30 that night, when most people who had not already left Herculaneum lived—briefly—to regret their delay. As Vesuvius tapped lower levels of its subterranean magma chamber, the gas-rich, volatile material that had sustained the eruption cloud became increasingly depleted, causing the immense ash column to collapse. Crashing to earth, the turbulent ash cloud expanded horizontally, separating into a fast-moving pyroclastic surge that roared through Herculaneum, killing the remaining people, and a somewhat slower-moving pyroclastic flow that enveloped the town, initiating its burial. As the eruption column continued to fluctuate wildly, like a hellish fountain gone crazy, a second pyroclastic flow was generated an hour later. A third, at 5:30 a.m. on August 25, finished the job of entombing Herculaneum under a shroud of seething lava fragments.

The third pyroclastic flow, which had incinerated Herculaneum, was deflected from Pompeii by the city's massive walls. The fourth, however, exhibited demonic energy, racing six miles from Vesuvius's crater and blasting through the city like a red-hot sandstorm. Sweeping across the nine-foot thickness of pumice that had previously fallen, the flow toppled walls, sheared off exposed roofs and upper stories of buildings, and asphyxiated the remaining inhabitants.

Caught in swirling masses of ash, thousands of fleeing Pompeiians were instantly entombed. When their bodies decayed, they left behind hollow spaces in the ash. Recognizing that these hollows were, in effect, molds of human figures, 19th-century archaeologists filled them with liquid plaster, allowed it to solidify, and then removed the surrounding ash. The plaster casts thus created were amazingly lifelike, some showing in detail the agonized facial expressions of the volcano's victims.

In a location called the Fugitives Garden, lie casts made in 1961 of seven adults and six children. Like many of the people trapped in the pyroclastic surge from Mount St. Helens in 1980, they appear to have died of suffocation from inhaling the pervasive ash. Estimates of how many people died in Pompeii—some archaeologists have said about 2,000—were based on the number of skeletons found in the lower pumice deposits and the human-shaped hollows preserved in the upper pyroclastic flow deposits. With the realization that the pyroclastic flows and surges probably killed simultaneously many of the people who were fleeing along the roads outside Pompeii's city gates, historians now believe that many more thousands perished in the catastrophe.

At 8:30 a.m., August 25, the younger Pliny and his mother witnessed the collapse of the sixth and final column, triggering the eruption's most terrifying pyroclastic surge, "a fearful black cloud…rent by forked and quivering bursts of flame" that swept across the bay toward Misenum. Along with hordes of others, they fled the city. As the monstrous cloud descended, "many besought the aid of the gods, but still more imagined there were no gods left, and that the universe was plunged into eternal darkness for evermore."

When the hurricanes of ash eventually dissipated and a murky daylight returned, Pliny found the region changed beyond recognition,

Asphyxiated by one of the pyroclastic flows that swept Pompeii and Herculaneum in August of A.D. 79, a Roman who failed to leave in time died in his doomed city. By pouring plaster into hollows found in the flow deposits, scientists preserved the shapes of many agonized victims.

"buried deep in ashes like snowdrifts." Although a few roofless walls and part of the large amphitheater, built atop a low hill, stood above the ashen waste, Pompeii lay smothered under 9 feet of pumice and another 6 to 10 feet of pyroclastic flow deposits. Herculaneum, sealed under pyroclastic deposits more than 65 feet deep in some spots, had completely vanished. The sites of the two cities remained a mystery until 1709, when laborers digging a well uncovered statuary that had fallen on the stage of Herculaneum's theater.

Although it was rediscovered first, most of Herculaneum still lies underground, embedded in pyroclastic deposits that hardened to the consistency of cement. Early excavators bored a network of tunnels through the town, looting the artworks for private collections. One of the most promising discoveries was an elaborate house known as the Villa of the Papyri, which contained a library of 1,800 carbonized scrolls. Hopes that this almost miraculously preserved collection would yield literary masterpieces, such as a lost tragedy by Sophocles, were dashed when most of it turned out to be the work of a minor Epicurean philosopher, Philodemus. The villa's presumed owner,

L. Calpurnius Piso, the father-in-law of Julius Caesar, was Philodemus's patron.

In the early 1980s, a grisly discovery of 150 skeletons in chambers lining Herculaneum's former beachfront gave insight into the town's inhabitants. The skeletons belong to people trying to escape by boat as fiery avalanches engulfed them. Among them were the bones of slaves and aristocratic ladies, whose remains still wore gold jewelry. All perished together inside what must have seemed a blazing furnace.

In the centuries following its destruction of Pompeii and Herculaneum, Vesuvius erupted intermittently, causing ash to fall as far east as Constantinople in A.D. 472. Shortly after the turn of the second millennium, the volcano entered a period of repose, which ended explosively in December, 1631, when 4,000 people were killed by swift-moving pyroclastic flows. That outburst, the worst since A.D. 79, introduced an eruptive pattern that prevailed well into this century. After a major eruption, Vesuvius would fall silent for several years, then begin mild activity that built a new cone filling the summit crater, only to have it blown away during a climactic explosion. This cycle, which characterized the volcano's behavior after 1631, ended with its eruption in March 1944. Since then, Vesuvius has remained ominously quiet, merely leaking steam from fumaroles in the summit crater.

VOLCANOES / MOUNT ST. HELENS

Located in southwest Washington State, Mount St. Helens is one of the youngest volcanoes in the Cascade Range. Rising about 6,000 feet above Spirit Lake at its northern foot, St. Helens's visible cone was constructed in the last 2,000 years, its smooth contours contrasting with the deep erosional scars Ice Age glaciers carved into its older neighbors, such as Mounts Rainier, Adams, and Hood.

Mount St. Helens's picture-postcard tranquility ended abruptly on March 20, 1980, however, when a series of small earthquakes began shaking the volcano. On March 27, an explosion ripped open a new crater in the summit ice cap, shooting steam and ash thousands of feet into the air. For the next six weeks, St. Helens intermittently spewed ash columns composed of pulverized rock from the volcano's summit dome. This initial activity, consisting entirely of phreatic (steam-blast) eruptions, was triggered by groundwater that flashed into steam when heated by magma rising through the volcano's internal plumbing system.

More ominous than the phreatic bursts was the swelling of St. Helens's north flank. Viscous magma injected into the cone had deformed the surface above it. By mid-May the bulge had pushed parts of the north slope more than 450 feet higher, setting the stage for a chain of cataclysmic events that changed the mountain and the surrounding region beyond recognition.

At 8:32 a.m., May 18, a Richter magnitude 5.1 quake, centered beneath Mount St. Helens's north slope, triggered one of the largest landslides in recorded history. The entire north side of the volcano suddenly peeled away, forming a massive avalanche about seven-tenths of a

JOHN MARSHALL (ALL)

cubic mile in volume. Rushing downslope at speeds up to 180 miles an hour, part of the avalanche surged into and across Spirit Lake, raising the floor of the lake by 295 feet and the water level by 200 feet. The main body of the avalanche, however, turned westward down the North Fork of the Toutle River, traveling more than 13 miles downstream and filling the valley to a maximum depth of 600 feet.

The sudden removal of Mount St. Helens's north side was like uncorking a bottle of warm, vigorously shaken champagne. The landslide uncapped the superheated water and gases dissolved in the body of magma that had entered the cone, unleashing a laterally directed pyroclastic surge that swept northward, devastating an area of about 230 square miles in a fan-shaped arc north of the volcano. A deadly mixture of incandescent gas, shredded magma, and rock fragments torn from the mountain, the surge raced outward at a velocity approaching

Viewed from the densely forested shores of Spirit Lake in April 1980, Mount St. Helens steams vigorously (left). The cataclysmic eruption of May 18th removed much of the volcano's north side, reduced its height by 1,300 feet, and littered Spirit Lake with timber flattened by a pyroclastic surge (center). Sixteen years later, a few pioneering plants restore some greenery to the 230-square-mile area of devastation.

the speed of sound (about 735 miles an hour), easily sweeping over ridges 1,500 feet high. Thousands of animals and 57 people perished.

Within 8 miles from the crater, the surge obliterated everything. In the zone between 8 and 19 miles from Mount St. Helens, Douglas firs up to 200 feet tall were immediately flattened. Along the outermost fringes of this "blow-down" zone was a thin line of seared trees that remained upright—the "standing dead." While the pyroclastic surge moved horizontally, a column of ash soared 90,000 feet into the stratosphere.

With the volcano's north side gone, fountains of magma jetting from the vent formed pyroclastic flows that swept northward toward Spirit Lake, setting the lake water boiling. At least 17 separate pyroclastic flows, seething masses of incandescent pumice, were emitted during the May 18 eruption. Two weeks later, some of the flow deposits still registered 785°F.

Second only to the avalanche and pyroclastic surge in destructiveness were the mudflows that poured down almost every stream valley heading on St. Helens. Mobilized by melting ice fields and river water, the largest flow swept down the North Fork of the Toutle River, coursed through the Toutle River Valley, then emptied more than 65 million cubic yards of sediment into the Cowlitz and Columbia Rivers.

Five smaller explosive eruptions occurred during 1980, sending ash plumes nearly ten miles into the air. Except for minor bursts in 1982, 1984, and 1990, most of St. Helens's subsequent activity was confined to quiet extrusions of stiff, viscous lava inside the crater. Scientists estimate that if the volcano resumes dome growth at its average rate of the 1980s, it would take nearly a century to fill the huge crater formed in the May 18 outburst and more than 200 years to restore Mount St. Helens to its pre-1980 height of 9,677 feet above sea level.

VOLCANOES / NEVADO DEL RUIZ

Human response to a volcanic crisis is literally a matter of life or death. When a long-dormant volcano near a populated area revives, local officials face a painful dilemma. If they decide to evacuate thousands of people from the vicinity and only a small, harmless eruption occurs, they are held responsible for unnecessarily frightening people and disrupting the area's economy. If the authorities ignore the threat and the volcano erupts violently, destroying a city and killing thousands of residents, they are held accountable for the catastrophe.

The 1985 eruption of Nevado del Ruiz, a large, glacier-clad volcano in Colombia, South America, killed an estimated 23,000 people—a loss that prompt government action could have prevented. This event was the worst volcanic disaster since Mount Pelée annihilated St. Pierre in 1902. Like most volcanoes, Nevado del Ruiz played fair, giving adequate warning of an impending eruption. Swarms of small earthquakes, caused by subterranean fracturing of rock as magma moved toward the surface, were recorded on seismographs at a station located about six miles from the volcano's summit. Scientists monitoring Nevado del Ruiz transmitted information about the increasing activity to Colombian emergency-response coordinators charged with alerting the public to the danger.

In 1984, when Nevado del Ruiz first began to shake and spew ash, Colombian geologists started mapping eruptive deposits around the volcano to assess the kinds of hazards it posed. They found that stream valleys draining the volcano's flanks had been repeatedly inundated in the recent geologic past by large-volume lahars, chaotic mixtures of water and rock debris that can travel downslope at speeds of 25 to 35 miles an hour. As recently as 1845, a lahar overwhelmed the town of Ambalema, about 20 miles southeast of Armero, killing 1,000 people. Unfortunately, the map delineating such potential hazards took nearly a year to complete and was available for distribution only days before the fatal eruption. Local authorities chose to disregard its implications and the recommendations of scientists.

What happened at Nevado del Ruiz the night of November 13, 1985, while the peak was veiled in storm clouds, has global significance because eruptions of precisely this nature have occurred—and will occur again—at similar volcanoes all over the world, from Japan to Russia to the western United States. A sudden explosion ejected hot rock fragments from the crater, triggering a pyroclastic flow that swept across the ice fields at the summit, almost instantly melting a large quantity of snow and ice. As heated floods of water poured into the canyons leading away from the summit, they picked up loose rubble from the valley walls and floors, creating hot lahars that grew in volume as they incorporated additional material lower on the volcano's slopes. Some of the lahars were nearly 150 feet thick.

Only two and a half hours after the eruption started, massive lahars reached the town of Armero, about 30 miles from the crater. The torrents of mud, boulders, and shattered trees took only a few minutes to race through Armero, burying most of the town and killing 20,000 of its inhabitants. Up to 3,000 more people died in other valleys that headed on the mountain.

Only a few of the roughly 4,200 buildings that once stood in the town of Armero, Colombia, rise above the thick lahars that poured from Nevado del Ruiz (above). In a terrible ordeal, rescuers struggled in vain to free a 13-year-old girl buried up to her neck in mud (left).

Rosa Maria Henao, whose home stood on a hill, was one of the fortunate survivors. She recalled that "the air suddenly seemed heavy. It smelled of sulfur. Then there was a horrible rumbling that seemed to come from deep inside the earth." Radiating an infernal scent of brimstone, the deadly morass of mud and debris "rolled into town with a moaning sound, like some sort of monster." Watching with her children from the roof of their house as most of the 4,200 other buildings in Armero disappeared forever, Señora Henao felt as if it were "the end of the world."

To prevent comparable losses in the future, the U.S. Geological Survey in 1986 created the Volcano Disaster Assistance Program (VDAP). Its goals were to help countries facing volcanic emergencies to reduce the number of potential fatalities and minimize economic disruption.

197

VOLCANOES /
MOUNT PINATUBO

Strategies developed by the Volcano Disaster Assistance Program (VDAP) for mitigating volcanic hazards were put to the test in 1991 when Mount Pinatubo, an obscure peak on the Philippine island of Luzon, suddenly gave signs of awakening from a long sleep. On April 2, Pinatubo began a series of steam blasts that ripped open a mile-long chain of vents high on the volcano's northwest flank. Ten weeks later this preliminary activity culminated in the world's most violently explosive eruption in more than half a century.

The news media focused on Pinatubo because it threatened two important U.S. military installations: Clark Air Base, which was located only 16 miles from the volcano's eastern foot, and Subic Bay Naval Station, about 26 miles to the southwest. Together, the two bases represented the centerpiece of the U.S. presence in Asia since World War II, and were the largest American military facilities abroad.

Almost immediately after the first steam explosions began, the Philippine Institute of Volcanology and Seismology (PHIVOLCS) set up several portable seismographs near Pinatubo's northwest base to record the hundreds of small earthquakes caused by magma rising into the cone. At about the same time, a third of the 5,000 people living on the volcano's slopes were evacuated beyond a radius of 6 miles from the summit. In late April, the

Filling the entire horizon on June 15, 1991, searing clouds of ash-laden gas—a lethal pyroclastic surge—broil from Mount Pinatubo in one of the 20th century's most violently explosive eruptions. Racing along a narrow dirt road, several people got away just in time.

Philippine scientists were joined by a group of volcanologists from the U.S. Geological Survey, and the international team established temporary headquarters at Clark Air Base, christened the Pinatubo Volcano Observatory (PVO).

Determined to avoid a repeat of the Armero disaster, the PVO scientists quickly studied deposits surrounding the volcano and prepared a hazards map pinpointing the areas at highest risk. They found that Pinatubo had left abundant evidence of a violent past, characteristically sending pyroclastic flows and surges down its flanks, mantling them with layers of tephra tens of feet thick.

Instruments monitoring Pinatubo's behavior provided strong clues that an eruption was imminent. Using their hazard map and analyses of the volcano's restlessness, the PVO scientists successfully persuaded the civil-defense officials and military commanders that Mount Pinatubo was extremely dangerous.

On June 10, a motorcade carrying 14,500 military personnel and their dependents left Clark Air Base, traveling to Subic Bay Naval Station, from which most flew back to the United States. In a decision that saved the U.S. government hundreds of millions of dollars, all remaining military aircraft, except for three helicopters, were also removed from the base.

By June 12, when the first in a series of large explosive eruptions began, tens of thousands of people had been evacuated beyond Pinatubo's lethal reach, including most of the remaining staff at Clark Air Base. The climactic outburst on June 15 ejected an enormous tephra column into the stratosphere, where it unfurled like a huge umbrella about 125 miles in diameter. By that date all but one of the seismometers installed along Pinatubo's base had ceased functioning, obliterated by the pyroclastic flows and surges that rushed down the volcano's slopes, choking valleys in thick deposits of ash and pumice. Some flows traveled as far as 10 miles from the summit, in places accumulating to depths of nearly 700 feet.

Falling tephra covered some 1,500 square miles around Pinatubo with an ash layer at least two inches thick. The destructive effects of the eruption were magnified by the unwelcome arrival of Typhoon Yunya, whose drenching rains saturated the ash accumulating on buildings, causing roofs to collapse and accounting for many of Pinatubo's nearly 1,000 fatalities.

Typhoon Yunya's downpour also mobilized the newly erupted pyroclastic deposits, transforming them into hot lahars streaming down river valleys. Even after the eruption had abated, each heavy rain generated floods and lahars that buried villages and farms, destroying houses and bridges and blocking numerous roads. In July, when the monsoon season began, economic losses rapidly increased as torrential rains washed even larger volumes of ash off Pinatubo's unstable slopes and inundated vast areas of valuable agricultural land. For several years after the eruption, lahars continued to form, submerging additional villages and cultivated fields.

The Pinatubo eruption also ejected a large quantity of tiny sulfuric-acid aerosol particles into the atmosphere, particles that winds eventually dispersed around the globe, creating a distinct climatic impact. Forming a fine mist that extended in a broad stratospheric layer,

Steam plumes inside the caldera formed by Mount Pinatubo's 1991 eruption dwarf the lone figure of a scientist taking gas samples. A lake partly fills the caldera, from which pulverized material suddenly exploded, including an estimated 20 million tons of sulfur particles. Propelled 25 miles into the stratosphere, the particles blocked sunlight and caused temporary global cooling.

Pinatubo's aerosol particles acted as a shield to solar radiation, thereby cooling the earth's surface by approximately one degree during an 18-month period after the eruption.

Although it is impossible to prevent destruction of property near an active volcano, scientists working at the Volcano Observatory proved that cooperation between geologists monitoring the volcano and local authorities can save lives. Pinatubo's outburst would undoubtedly have killed thousands more had government agencies not agreed to evacuate the endangered area.

This outstanding success in mitigating the loss of life during a catastrophic event provides a model for future handling of volcanic crises.

VOLCANOES / HOT SPOTS

Although located more than 2,000 miles from the nearest plate boundary, the Big Island of Hawaii contains two of the world's most active volcanoes, Kilauea and Mauna Loa. In Hawaiian mythology, Pele, the volcanic fire goddess, now makes her home in Halemaumau, a deep pit inside Kilauea's summit caldera. When a restless spirit moves her, Madame Pele takes up temporary residence in Mokuaweoweo, the caldera indenting the summit of Mauna Loa (Long Mountain), or even the volcano's lower flanks. Wherever Pele camps, her presence is marked by fountains of molten rock jetting high into the air, producing streams of lava that flow like fiery, shimmering rivers into the sea, creating new land.

Local myth states that before she settled in Halemaumau, Pele ignited mountain-building fires on other islands. But the goddess was driven from her former abodes on Kauai, Oahu, and Maui by her sister Namaka, ruler of the sea.

These ancient traditions reflect geologic fact. Constructed during the past 700,000 years, the Big Island is the youngest in the Hawaiian chain and is still growing. Much of Mauna Loa's surface has been paved over by new lava flows during the past few thousand years. A full 90 percent of Kilauea's surface was formed during the past thousand years, some of it only yesterday. To the northwest, the islands are progressively older and consequently more eroded by the relentless battering of sea waves, the destructive work of Namaka.

An observant traveler heading northwestward along the 1,500-mile-long Hawaiian archipelago to the island of Midway will find that Namaka's ultimate triumph is complete. Erosion has worn the northernmost volcanic islands to below sea level, allowing coral atolls to form atop the drowned mountains. Midway, millions of years older than Hawaii, represents this late stage of the islands' geologic evolution. Farther north, a line of submarine volcanic cones, known as the Emperor Seamount chain, extends along the Pacific floor until it disappears into the deep trench between Kamchatka and the western tip of the Aleutian Islands.

In the rivalry between Pele and her sister, Namaka has time on her side. A restless wanderer, Pele typically finishes one project and then abandons it to the elements, moving on to the next. She completed Kauai almost six million years ago and Oahu three million years later, giving Namaka ample leisure to erode these islands. Namaka has had a full 70 million years to tear down the Emperor Seamounts, which once rose as high above the Pacific Ocean surface as Hawaii and Maui stand today.

Pele's mythical flight southward to her present residence in the Halemaumau fire pit offers an important clue to the geologic processes that created the Hawaii–Emperor Seamount chain. Like the Yellowstone plateau on the United States mainland, the Big Island rises above a hot spot rooted in the earth's mantle. A geographically isolated concentration of high heat energy,

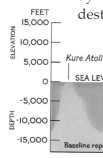

FEET
ELEVATION
15,000
10,000
5,000
0 SEA LEVEL
DEPTH
-5,000
-10,000
-15,000

Kure Atoll

Midway Islands
(27.7 million years old)

Pearl and
Hermes Atoll

Lisianski
Island

Maro Reef

Laysan
Island
(19.9 million
years old)

Gardner
Pinnacles

Baseline represents -18,000 feet, the average depth of ocean floor.

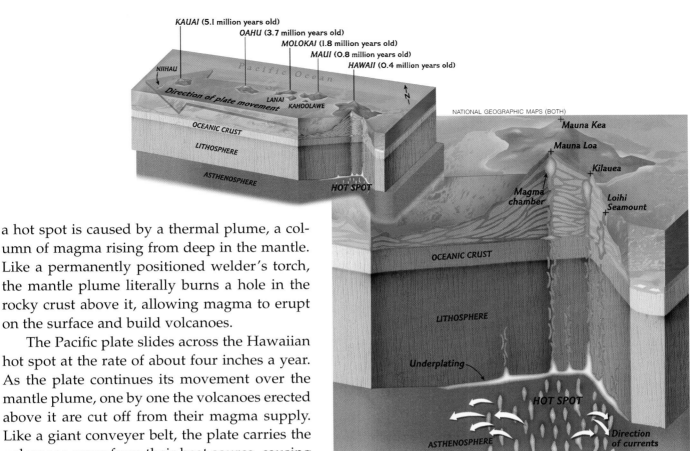

KAUAI (5.1 million years old)
OAHU (3.7 million years old)
MOLOKAI (1.8 million years old)
MAUI (0.8 million years old)
HAWAII (0.4 million years old)

NIIHAU
Pacific Ocean
LANAI
KAHOOLAWE
N
Direction of plate movement

OCEANIC CRUST
LITHOSPHERE
ASTHENOSPHERE
HOT SPOT

NATIONAL GEOGRAPHIC MAPS (BOTH)

Mauna Kea
Mauna Loa
Kilauea
Magma chamber
Loihi Seamount

OCEANIC CRUST
LITHOSPHERE
Underplating
HOT SPOT
ASTHENOSPHERE
Direction of currents
Magma pods

a hot spot is caused by a thermal plume, a column of magma rising from deep in the mantle. Like a permanently positioned welder's torch, the mantle plume literally burns a hole in the rocky crust above it, allowing magma to erupt on the surface and build volcanoes.

The Pacific plate slides across the Hawaiian hot spot at the rate of about four inches a year. As the plate continues its movement over the mantle plume, one by one the volcanoes erected above it are cut off from their magma supply. Like a giant conveyer belt, the plate carries the volcanoes away from their heat source, causing their fires to sputter and die. No longer replenished by infusions of magma, the dying volcanoes are defenseless against unending erosion by storm and stream. Namaka's abrasive agents cut deeply into the volcanoes' flanks and summit calderas, reducing their elevations and excavating enormous canyons like those seen today at Haleakala volcano on Maui.

Hawaii's volcanic landlords represent every stage of the construction-destruction cycle. Regularly erupting immense volumes of fluid basalt, Mauna Loa and Kilauea are vigorously building cones. Nearby Mauna Kea has entered a later stage typified by the sporadic eruption of more silicic, explosive magma, which has erected overlapping cinder cones and thick lava flows largely filling the summit caldera.

As the older islands drift northward to their final subduction in the Aleutian Trench, Pele is preparing a fresh addition to Hawaii's volcanic

family. About 20 miles south of Kilauea the Loihi Seamount is rising from the seafloor. Loihi is not expected to break the ocean's surface for 60,000 years; but when it does, it will announce its arrival with violent blasts of steam as hot lava mixes with shallow seawater. After a tempestuous introduction, Loihi will settle into the kind of familiar effusive eruptions we see today at Kilauea, constructing a new tropical paradise for our vacationing descendants.

Hawaii's chain of islands rose from the waves as the Pacific plate traveled across a hot spot in the mantle. Rising magma continues to build the Big Island (above). The moving plate carries older islands away from the magma source, allowing rain and waves to erode them (below).

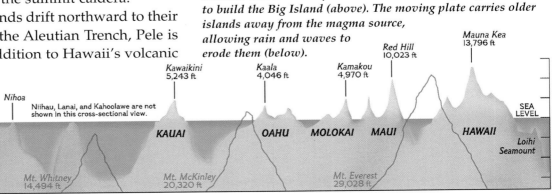

Mauna Kea 13,796 ft
Red Hill 10,023 ft
Kamakou 4,970 ft
Kaala 4,046 ft
Kawaikini 5,243 ft
Nihoa
Niihau, Lanai, and Kahoolawe are not shown in this cross-sectional view.

KAUAI   OAHU   MOLOKAI   MAUI   HAWAII
SEA LEVEL
Loihi Seamount

Necker Island (10.3 million years old)
rench igate oals

Elevations in red measured from base to summit.
Mt. Whitney 14,494 ft
Mt. McKinley 20,320 ft
Mt. Everest 29,028 ft

Yellowstone

Pele's mid-oceanic hot spot keeps a cauldron of liquid basalt on a comparatively gentle boil, but the mantle plume currently simmering under Yellowstone National Park concocts a more dangerously volatile brew. Yellowstone's hot springs, spouting geysers, and brilliantly colored mineral terraces, which draw millions of visitors every year, are but mild indicators of the powerful volcanic forces at work beneath the surface. America's first national park is the site of three of the most violently explosive eruptions known anywhere on earth. It may, geologists fear, be evolving toward a fourth.

The Yellowstone hot spot has a long history. Its path across western North America, many geologists believe, can be traced eastward from the Columbia River plateau, through Idaho's Snake River Plain, to its present location under the Yellowstone plateau in northwestern Wyoming. The hot spot's elongated track from west to east represents the westward drift of the North American plate as it traveled for millions of years over a thermal plume. The oldest evidence of the hot spot's presence are the floods of basaltic lava that inundated eastern Washington and Oregon, beginning in Miocene time about 17 million years ago. Issuing from fissures many miles in length, basalt swept over tens of thousands of square miles of the Pacific Northwest, with single flows traveling 300 miles down the Columbia River's ancient channel and emptying into the Pacific. No eruptions of historic time have come near to equaling the vast quantities of molten rock—60,000 cubic miles—that poured out, mostly during a relatively brief period of two or three million years.

PAUL CHESLEY (BOTH)

Some geologists speculate that these basaltic flood eruptions were triggered by the impact of a meteor or asteroid, such as that which apparently initiated the even more voluminous Deccan Plateau flow in western India.

As the tectonic plate carrying North America slipped over the hot spot, the basaltic eruptions in Washington and Oregon gradually ceased and a series of violently explosive eruptions began to the east. When the solid granitic crust of southern Idaho slowly passed over the mantle plume, pockets of granitic rock were melted to form a highly silicic magma known as rhyolite. Beginning about 13 million years ago in southwestern Idaho, rhyolitic magma was ejected in a prolonged sequence of huge pyroclastic flows and surges that left thick deposits of light-colored ash smothering the state. Fallout from the towering ash clouds deposited thick layers of ash east of the Rockies and over the northern Plains states.

About 2.2 million years ago, after the North American plate had moved farther west, eruptions began in the Yellowstone region. The hot spot melted the granitic crust, forming a large gas-rich body of subsurface rhyolite magma. By 2 million years ago, enough magma had accumulated to produce one of the largest explosive outbursts to occur on earth. The most voluminous of the three Yellowstone eruptions, this

Yellowstone National Park preserves one of the world's most spectacular thermal areas. A subterranean hot spot fuels thousands of geysers, springs, and steam vents, including Upper Geyser Basin (right) and a large hot spring (opposite). North America's three greatest eruptions happened here.

event spewed out more than 600 cubic miles of fresh magma, enough to build 6 mountains the size of California's Mount Shasta.

Although no human has witnessed a volcanic event of this magnitude, geologists who have studied the deposits laid down and observed similar but smaller historic eruptions can guess what took place. Escaping gas whipped the magma to a glassy froth, which burst simultaneously from a series of concentric crustal fractures that had formed above the subterranean magma reservoir. After jetting to stratospheric heights, a massive column of gas and incandescent ash collapsed to spread outward in all directions. The huge waves of ash traveled enormous distances, sweeping over ridges and peaks and filling valleys with still-molten material. The clouds of ash surged over thousands of square miles so quickly that the incandescent rock lost little heat in transit. Deposited at extremely high temperatures, glassy fragments in the ash surges melted and fused together, welding the ash particles

together to form a solid rock called ignimbrite.

So much material was ejected that the roof of the underground magma chamber collapsed, causing the overlying crustal rocks to subside several thousand feet. The resulting caldera covered an area of about a thousand square miles. Extending across Island Park and the Yellowstone plateau, the caldera outlines have been largely obscured by later eruptive and erosive activity.

The second great explosive episode, which took place about 1.3 million years ago, largely duplicated the events of the first. The smallest of the catastrophic outbursts, this eruption ejected about 70 cubic miles of material and formed the Island Park caldera, about 17 miles in diameter. The third and latest eruptive cycle began about 1.2 million years ago, ejecting at least 250 cubic miles of seething magma fragments—approximately 1,000 times the volume of material that Mount St. Helens erupted in 1980. Although not as voluminous as the initial eruptions, the ash column reached high into the stratosphere and probably darkened the skies over much of North America. Geologists have found ash deposits in locations as distant as California, Kansas, and Saskatchewan, Canada. Rapid emptying of the magma chamber caused its roof to collapse, forming the present caldera, which measures about 30 miles across and 50 miles long—one of the largest volcanic depressions on the planet.

The Yellowstone volcano continued to erupt intermittently during the next several hundred thousand years, emitting an additional 250 cubic miles of material, mostly in the form of lava flows and domes. Although the most recent effusive eruptions occurred about 70,000 years ago, the hot spot maintains a large reservoir of magma not far below the surface, providing enough heat to keep some 10,000 hot pools and springs, geysers such as Old Faithful, and pots of bubbling mud on the boil. The continuing thermal activity, intermittent earthquakes, and measurable uplift of the caldera's floor demonstrate that the Yellowstone area remains geologically restless. Whether the hot spot is generating a new batch of rhyolite magma to fuel another cataclysmic eruption, or whether it is at the end of an eruptive cycle, is not known.

The Yellowstone hot spot, which has persisted for at least 14 million years, will probably reach North Dakota and the Midwest eventually. Although many thousands of years may go by before another caldera-forming cataclysm occurs, it is sobering to reflect that the caldera at Yellowstone has the potential to explode on a scale that could transform much of North America into an ash-shrouded desert.

Iceland

The heavily glaciated island of Iceland, with an area of roughly 40,000 square miles, owes its existence to a hot spot on the northernmost extension of the Mid-Atlantic Ridge. Along with the Azores and Ascension Island, Iceland represents one of the few places in which volcanoes along the submarine ridge rise high enough to emerge above the ocean surface. Built astride an active spreading center, Iceland displays a textbook example of the process by which plates diverge and separate. A deep trough, which characterizes the underwater summit of the Mid-Atlantic Ridge, also cuts through Iceland's land surface, a feature demonstrating that the island is being torn apart as plates on both sides of the ridge pull away from each other. Magma rising through a wide zone of crustal fractures, with a generally northeast-southwest axis, continues to split Iceland in two, giving geologists an opportunity to study an aspect of plate tectonics that is ordinarily hidden beneath the sea.

Like Hawaiian and other mid-oceanic volcanoes, Iceland's volcanoes produce comparatively gentle effusions of basaltic lava. Following the arrival of Viking settlers in A.D. 874, eruptions have occurred, on average, about every five or six years. Although some eruptions have built shield structures resembling those in Hawaii, the most distinctive kind of Icelandic activity is the fissure eruption, such as that of Mount Laki in 1783. The most voluminous outpouring of lava in historic time, far larger than that of any Hawaiian volcano, the Laki eruption began when a series of fissures 15 miles long suddenly split the earth near Iceland's southern coast.

Steam issues from vents melted in Iceland's Vatnajökull, whose 3,200 square miles make it the largest glacier in Europe. A battleground of glacial ice and volcanic fire, Iceland sits astride a spreading center on the Mid-Atlantic Ridge. In the fall of 1996, an eruption beneath its vast ice cap generated floods of meltwater, inundating lowlands and destroying roads and bridges.

After initial fountains of molten rock and mild explosions hurled tephra into the air, immense floods of extremely fluid basalt burst from the fissures, streaming down river valleys and fanning out in a broad fiery delta on the coast. Within two months, from early June to August, these copious flows had buried 226 square miles of valuable pastureland and river bottoms under jagged flows of basalt.

Gases released during the Laki eruption were even more destructive to plant and animal life. Forming a bluish haze, probably of sulfur dioxide, a gaseous miasma hovered over the landscape through the summer of 1784, blighting vegetation and stunting crops. In the resulting "haze famine," Iceland lost three-quarters of its sheep and horses, one-half of its cattle, and one-fifth of its 49,000 people, a disaster from which it took years to recover.

Enough ash and sulfur particles were injected into the atmosphere to cause a similar haze in other parts of the northern hemisphere, including the eastern United States, where Benjamin Franklin was apparently one of the first scientists to speculate that distant volcanic eruptions could both pollute the air and promote climatic cooling.

Other notable Icelandic eruptions include the 1947-48 Mount Hekla event, which ejected a pall of ash over Finland some 500 miles away, and the 1963 eruption of Surtsey, a volcano that rose above the ocean's surface and created a new island off Iceland's southern coast.

Erupting explosively when it first broke the surface, Surtsey initially built a low cone of fragmental material, which was immediately attacked by waves that threatened to eradicate the entire island. In 1964, however, Surtsey guaranteed its continued existence by emitting fluid streams of lava that soon mantled the cone in a hard protective shell. After three and a half years of vigorous activity, Surtsey fell silent in 1967. Only six years later, on the nearby island of Heimaey, yet another Icelandic volcano suddenly appeared, testimony to the continuing vigor of this Mid-Atlantic hot spot.

VOLCANOES / BENEFITS

Volcanoes can be agents of destruction, but in the long run they are also creators and benefactors. Lava flowing into the sea builds new land, increasing the areas that will eventually become habitable.

Volcanic soils are famous for their fertility, accounting for prosperous vineyards, farms, and grainfields from Italy to Indonesia. On the island of Java in the Indonesian archipelago, where volcanoes frequently spew ash over the countryside, farmers routinely produce two or three rice crops a year, far more abundant yields than those on Borneo, which has no volcanoes. Because ash is rich in potassium and phosphorus, two elements necessary for healthy plants, its presence enhances productivity. Whereas a heavy ashfall can kill vegetation, adding a thin layer of fresh ash seems to stimulate growth.

Volcanic heat energizes most of the world's hot springs and thermal areas, many of which are highly valued for their therapeutic benefits. When heat from subterranean magma encounters groundwater, the water—constrained by the weight and pressure of overlying rock—may reach 500°F without boiling. Lighter and less dense than the cool water percolating down from above, the superheated water rises to the surface along fractures in the crustal rock to form a hot spring or steam vent.

Italy, Mexico, Iceland, Japan, New Zealand, El Salvador, and the United States have developed technologies that use geothermal energy to produce electric power. Steam or superheated water in porous layers of volcanic rock is piped out of the ground to run turbines that generate electricity. Because geothermal energy

Miners rest on chunks of sulfur in an active volcanic crater on Java. In mining this nonmetallic element—used to make such products as medicines, paper, and gunpowder—workers fight off nausea caused by inhaling its rotten-egg odor. To mitigate their discomfort, the miners labor on an empty stomach. Companies seek valuable sulfur deposits around the world, from Indonesia's volcanic peaks to Mexico's Popocatépetl.

production does not require the burning of fossil fuels, it is less environmentally polluting than most conventional plants.

Geothermal energy can also be used for nonelectrical purposes. About 85 percent of Iceland's houses are heated with water piped from underground. In the capital, Reykjavik, almost every dwelling draws on volcanic heat to provide its occupants with warm rooms and hot showers.

The world's largest geothermal facility sprawls over 15 square miles of hills at The Geysers, about 90 miles north of San Francisco, California. Because this field taps directly into high-pressure steam rather than hot water, the problem of separating steam from water does not exist, making electrical power easier and more economical to produce. In recent years, The Geysers increased its production capacity to 2,000 megawatts of electricity—enough to supply 2 million people—but production has since declined. Some geologists suspect that the field may literally be running out of steam.

Volcanic action is also responsible for creating much of the earth's mineral wealth. As subterranean magma bodies slowly cool and crystallize, the process releases minerals dissolved in the magma. Hot water circulating underground leaches minerals from the magma and deposits them in cracks of crustal rock, forming veins of gold, silver, copper, zinc, and lead.

Because of plate movement and subduction, some minerals deposited on ancient seafloors are now accessible to mining on the land surface. Convergence of the African plate with the Eurasian plate not only generated the belt of Mediterranean volcanoes but also created the rich copper deposits on Cyprus. As the African plate descended, an ore-rich slab was thrust against Cyprus, whose mines provided copper for the tools and weapons of early civilizations.

Diamonds, among the rarest and most precious of gems, are created by a complex volcanic process that even now is not completely understood. Unlike mineral ores formed by hydrothermal action, diamonds are forged from carbon atoms under enormous pressures and at great depths. After formation, they are rapidly transported along volcanic conduits, called "diamond pipes," and are injected into near-surface deposits. After these deposits have been exhumed by erosion, the gems can be mined. Perhaps formed only during rare stages of earth's geologic history, diamonds are found today in eroded volcanic pipes in South Africa, Siberia, and Australia.

Within somewhat easier reach are sulfur deposits, such as those in the crater of Mexico's Popocatépetl. In the 16th century, Spanish conquistadores forced Aztec captives to climb the 17,000-foot volcano and descend into its crater to extract sulfur for making the gunpowder needed to complete Mexico's subjugation.

Volcanoes also have an incalculable aesthetic value. As Japanese artists have demonstrated, Fujiyama's exquisite symmetry offers a visible image of the intersection of heaven and earth, its cloud-piercing summit suggesting dimensions far beyond mere physical existence. Profoundly aware of volcanic mountains' power to inspire and heal, American naturalist John Muir urged people to seek spiritual renewal in the presence of such sleeping giants as Mounts Rainier and Shasta.

VOLCANOES / HAZARDS

Looming above Puget Sound, Mount Rainier is the highest and perhaps the most dangerous peak in the Cascade Range. Deceptively quiet since a minor eruption in the mid-19th century, Rainier has the capacity to devastate areas where more than 100,000 people live. During its last major eruptive period, between about one or two thousand years ago, lava from the summit crater partly melted the Emmons and Nisqually glaciers, generating floods and mudflows that swept down the White and Nisqually River valleys.

In the last 10,000 years, more than 60 such debris flows have streamed down valleys heading on the mountain. During an outburst 5,700 years ago, Rainier's summit collapsed, sending almost a cubic mile of rock into the White River Valley and forming an enormous mudflow that traveled 65 miles to inundate Puget lowlands. A repetition of that event today would partly or entirely destroy numerous cities and towns lying between the Cascades and Puget Sound.

South America leads the world in the number of volcanoes that threaten populated lowlands. Of the 204 volcanoes cataloged thus far in the Andes Range, 122 bear a heavy mantle of glacial ice, and many of them are capable of producing devastation comparable to the 1985 disaster in Armero, Colombia.

The Long Valley–Mono Lake region of east-central California is troubled by a persistent volcanic restlessness that began in May 1980. As the site of one of the greatest volcanic eruptions known, the area merits careful monitoring. About 700,000 years ago, the Long Valley volcano ejected about 140 cubic miles of pumice, causing pyroclastic flows energetic enough to

surmount the crest of the Sierra Nevada and race westward down the San Joaquin River drainage, perhaps as far as California's Central Valley. Geologists do not expect an imminent replay of that cataclysm, but smaller explosive eruptions along the Mono Lake–Inyo Craters chain of vents occur on the average of every 500 years, the last about 600 years ago. Several towns in the vicinity, including the winter resort of Mammoth Lakes, may be affected by eruptions in the future.

Beginning in 1989, a sequence of earthquakes centered near Mammoth Mountain, a massive cluster of lava domes banked against the Sierra's steep eastern wall. Sporadic quakes continue, along with emissions of carbon dioxide and helium gas that since 1990 have killed innumerable trees growing in this popular ski area. Even if magma that was recently injected beneath the mountain does not ultimately erupt on the surface, rising magmatic heat could trigger dangerous steam-blast eruptions similar to those that opened new craters on Mammoth Mountain about 700 years ago.

In Costa Rica, a small composite cone named Arenal produced its first historic outbreak in 1968. It suddenly bombarded adjacent towns and cultivated fields with a hail of lava bombs, some of which exploded on impact and excavated pits more than 160 feet in diameter. Following the initial explosions, Arenal has remained active, oozing viscous lava flows from vents on its flanks and expelling moderate amounts of ash—a persistent nuisance to settlements downwind.

Virtually every country bordering the Pacific Rim, including Russia, Japan, Chile,

Snowy Mammoth Mountain dominates the popular Sierra ski resort of Mammoth Lakes in east-central California. A cluster of massive lava domes, the mountain lies in a volcanic region known as Long Valley–Mono Lake, where geologic restlessness threatens an eruptive event. Earthquake swarms that began in May 1980 and emissions of heat and steam seem to warn that a volcanic eruption may occur soon.

Colombia, Nicaragua, Mexico, and the United States—as well as nations from Italy to Turkey and Kenya to Zaire—face a volcanic threat. The fact that in some regions no volcano has erupted during historic time offers little reassurance. Of the 16 largest explosive eruptions during the 19th and 20th centuries, which together killed some 140,000 people, 12 occurred at volcanoes with no known historical activity. (The fatalities do not include the 53,000 who died at Pelée and Nevado del Ruiz, which staged relatively small eruptions). In fact, the longer the interval is between eruptions of the world's composite volcanoes, the larger and more violent the next outburst is likely to be. Unfortunately, geologists have no way to determine which of the world's hundreds of sleeping fire-monsters is likely to revive soon.

LIVING ON UNSTABLE LANDS

We do not know how our remote forebears responded to volcanic threats, but people living all over the world, from the New Stone Age to the present, have shown incredible ingenuity in coping with these unpredictable agents of chaos.

The earliest known picture of a volcanic eruption, dating from about 6200 B.C., was found in excavations of Çatal Hüyük in central Turkey. The prehistoric artist who painted a wall mural showing a cinder cone spouting lava fragments obviously considered the activity significant, perhaps a manifestation of the power of the primordial Great Goddess whose enthroned image was also found in Çatal Hüyük's ruins. About 8,000 years later, in the fall of 1996, some Mexican villagers living near smoldering Mount Popocatépetl made pilgrimages to caves on the volcano's flanks, offering food and flowers to appease chthonic (underworld-oriented) forces not unlike those whom the ancient goddess figure represented.

With every passing year, more people are forced to deal with geologic violence. By the year 2000, approximately 500 million people will live on or near potentially dangerous volcanoes; several times that number already inhabit areas susceptible to destructive earthquakes. In addition to using time-honored religious or magical means to protect themselves, people living in seismically or volcanically hazardous zones can benefit from a variety of modern scientific techniques to mitigate losses of life, property, and industrial production.

The first step is to learn which areas are at highest risk and to define the precise nature and degree of the potential threat. In the case of volcanoes, regions with a long recorded history of eruptions, such as Japan, Iceland, and the Mediterranean area, have a good idea about the way in which individual volcanoes behave and can plan accordingly.

In countries where volcanoes erupt less frequently or where historical documentation covers only the last century or two, geologists must discover a volcano's potential threat by studying and mapping deposits left by prehistoric eruptions. By preparing maps that show not only the direction in which prevailing winds typically carry ashfalls, but also the location and length of lava flows, pyroclastic surges, and mudflows, geologists can pinpoint the areas most likely to be affected by a particular phenomenon during future activity.

In compiling a map for Mount Pinatubo before its 1991 outburst, geologists were able to inform authorities that in the recent geologic past the volcano had erupted pyroclastic flows extending miles beyond its summit, a finding that helped officials to realize the potential danger they faced. In 1978, two scientists with the U.S. Geological Survey, D. R. Crandall and Donal Mullineaux, published a now-classic report detailing hazard zones around Mount St. Helens and warning that the volcano was likely to erupt again soon, "perhaps before the end of this century"—a forecast that came true with a vengeance only two years later.

By dating volcanic deposits, geologists can determine how often a volcano erupts and compute the likelihood of an eruption occurring in any given year. As yet, however, scientists can not predict when a now-dormant volcano will awaken or how large its next eruption will be.

One of the earth's most active volcanic regions, the island nation of Indonesia fosters a culture that shows profound respect for its 76 historically active fire-mountains. A group of prayerful Indonesians— hoping to appease the potentially destructive spirits believed to inhabit Java's Mount Bromo—has gathered at the rim of the volcanic crater to make offerings of money, food, and beverages.

PHILIP JONES GRIFFITHS/MAGNUM

Nor, in the case of composite volcanoes, which typically exhibit a variety of eruptive styles, can scientists forecast the exact nature of the eruption. Vesuvius and Mount St. Helens, for example, have an extensive repertory, ranging from quiet outpourings of fluid lava to mildly explosive activity to extremely violent outbursts.

Once a previously quiet volcano begins to stir, geologists employ several different instruments to monitor its activity, helping them to determine how soon it will erupt fresh magma. Seismometers placed at strategic locations around the volcano record earthquakes generated by magma or other fluids moving underground. Tiltmeters record ground deformation, such as a swelling of the volcano as magma rises through the volcano's interior. Laser beams bouncing off reflectors placed on the volcano's slopes also detect slope displacement caused by magma infusion. Additional devices are used to register temperature changes and increasing gas emissions.

Armed with such measurements, scientists can then alert officials who must then make crucial decisions about evacuating tens of thousands of people and prohibiting access to the endangered area. Worldwide, most government officials seem to find making such decisions difficult, primarily because they can disrupt businesses and possibly engender economic losses. Despite the social upheaval it would cause, however, in June 1991 U.S. military and Philippine civil authorities agreed to order an evacuation from the hazard zone around Mount Pinatubo. This action saved thousands of people from being incinerated in pyroclastic flows that soon afterward swept down the volcano's slopes.

Well before a quiescent volcano revives, officials can minimize future damage by using hazard maps in long-range planning, making sure that new developments are not built in areas of highest risk. Floors of valleys heading on high, steep, glacier-covered cones—such as the volcanoes in the Andes of South America and in the Cascade Range of the western United States—are particularly vulnerable to large-scale avalanches, mudflows, and floods. No matter how scenically attractive, valleys like the one where Armero stood are not safe locations in which to build population centers.

Whereas an average of 60 volcanoes erupt each year, 100,000 earthquakes strong enough to be felt shake the planet annually. Of these, no fewer than 1,000 are of sufficient size to cause damage. In the mortality sweepstakes, earthquakes clearly outclass volcanoes. During the past 1,000 years, eruptions have killed about 300,000 people. Single earthquakes, however, have claimed more than twice that number of lives in a matter of minutes. Perhaps history's worst natural disaster took place in 1556, when 830,000 perished in an earthquake in China's Shaanxi Province. China suffered an almost equally high death toll in the 1976 Tangshan quake. Although official government estimates put the number of deaths at about 240,000, outside observers estimate that perhaps as many as 655,000 to 800,000 people died at Tangshan.

The capacity of earthquakes to wreak such havoc is a compelling inducement to find ways to reduce their toll. Although the Kobe earthquake in 1995 demonstrated the fragility of

Erected on shaky ground, storage tanks on Tokyo Bay (opposite) may suffer major damage when the next great quake rattles this part of Japan and turns water-saturated landfill into quicksand. Many of the port facilities encircling the Pacific basin sit on vulnerable bay mud. While Sakurajima spews ash in the background, Japanese schoolchildren wear protective helmets to ward off falling tephra (left).

man-made structures when large events center near urban areas, many scientists believe that there are effective means of mitigating losses in future quakes.

To identify zones of high risk, geologists study an area's earthquake history, looking for previously unrecognized faults, as well as faults known to have produced damaging shocks in historic time. Some intraplate faults, like many in China and other parts of the Eurasian continent, are buried so deeply beneath sediment that their existence remains unsuspected until they trigger a major earthquake.

Utilizing the experience of how different kinds of soil behave during strong shaking, geologists can compile seismic hazard maps, showing exactly which areas of Tokyo, Yokohama, San Francisco, Los Angeles, or Mexico City will suffer the highest Mercalli intensities. Knowing which places are the most susceptible to liquefaction or subsidence, engineers can strengthen older structures, retrofit overpasses, and construct buildings designed to survive ground failure.

New quake-resistant techniques include the concept of seismic isolation, in which the base of a building or highway lies atop rubber-and-steel pads. The pads act like springs or shock absorbers, reducing the degree of ground motion transmitted to the structure. In the Northridge quake, a seismically isolated hospital remained virtually untouched, while a conventionally built medical center next door sustained 389 million dollars in damage. In the Kobe quake, two buildings erected on the seismic isolation principle reportedly escaped unscathed.

Using another promising innovation, engineers design special steel configurations, lead shock absorbers and similar dampers, to reduce an edifice's swaying motion. Enforcing strict building codes and extensive retrofitting of existing structures will be enormously expensive, but—as many scientists point out—much less costly than enduring the loss of billions of dollars in future quakes.

In areas from New England and New York to Portugal and Russia, where large quakes are spaced far apart in time, complacency allows substandard construction that will probably fail catastrophically, reminding us that only intelligent action can limit the extent of human suffering when the planet shrugs its brittle skin.

CLIMATIC SHIFTS
OMENS OF
GLOBAL WARMING?

by Michael H. Glantz

INTRODUCTION

When scientists refer to the climate system, most people think they are talking only about rainfall and temperature. But the climate system is much more than that, encompassing snow and ice, vegetation and soils, oceans and lakes, volcanoes, and so on. Recently, governments have defined the system solely in terms of the natural environment. The truth of the matter is, however, that we cannot really understand the workings of the climate system if we do not take into account the effects human activities have on it.

The atmosphere is a receptacle for dust from the desert regions, as well as for gases and particulates emitted by natural factors such as volcanic eruptions and by human factors that include industrial and land-use activities. These gases and particulates can affect atmospheric processes such as heating and cooling, cloud formation, and winds. The atmosphere has been referred to as the 'fast' climate system because its 'memory' is relatively short.

The oceans are an integral part of the climate system. Their surface waters and the atmosphere constantly interact, with one either heating or cooling the other, thereby greatly influencing global and regional climate conditions. Referred to by climatologists as the 'slow' climate system, oceans provide the global system with a relatively long memory—on the order of centuries. Deep cold ocean water, for example, may take 1,500 years to turn over (come to the surface).

Snow and ice on land and in the sea are extremely important to the climate system because they reflect much of the incoming radiation from the sun. The characteristic of snow and ice involving reflectivity is what scientists call the albedo effect. Different parts of the earth's surface reflect the sun's rays to varying degrees: Darker surfaces absorb radiation more than lighter ones do. The amount of radiation absorbed or reflected by the surface directly affects the degree of atmospheric heating.

The earth's surface interacts with the atmosphere in a variety of ways. Vegetation transpires, putting moisture into the atmosphere; plants pull carbon out of the atmosphere as they grow and put it back when they decay or burn. Vegetation also releases nitrogen, sulfur, and phosphorus into the air as part of what are called the biogeochemical cycles. A report by the National Aeronautics and Space Administration (NASA) has noted that the movement of these key elements through the earth system is critical to the maintenance of life.

In the mid-1970s, professor Jule Charney of the Massachusetts Institute of Technology suggested that the devastating drought in the West African Sahel in the 1960s and '70s resulted from too many cattle eating too much vegetation. When the vegetative cover is removed, soils that have a high albedo are exposed to the sun's rays. The increase in reflectance brings stability to the air above the land, cooling the surface and thereby inhibiting the atmospheric processes that produce rain.

In addition to its ability to absorb solar radiation and pull carbon from the atmosphere, vegetation performs other functions in the climate system. A good example lies in the Amazon rain forest of Brazil. Much of the rain falling in the Amazon basin is moisture evaporated from rain forest vegetation. More specifically, half of the region's rainfall comes from within the basin,

Darker areas such as tropical rain forests (left, lower) absorb more of the sun's rays than do lighter ones such as polar regions (upper). The amount of solar radiation absorbed or reflected by the planet's surface affects the heating of the atmosphere. Forests also pull carbon dioxide out of the atmosphere and store it. Cutting down or burning the trees releases the stored carbon, enhancing the greenhouse effect.

PRECEDING PAGES: At the desert's edge in Niger, herds feed on vegetation that protects fragile soils.

producing acid rain on the regional scale and climate change on the global scale. In Central Asia, withdrawals from feeder rivers are shrinking the Aral Sea and changing the region's climate. In Africa and elsewhere, people collecting firewood to meet essential energy needs are denuding the land, leaving it vulnerable to wind erosion and a subsequent drop in soil fertility.

People have also sought to combat climate's adverse aspects. Some governments have built seawalls or dams to protect settlements from flooding. Others have grown walls of trees to weaken hot, dry winds or have banned the use of chemicals that destroy the ozone layer.

Many recent reports from the research community suggest that human activities are at the root of the climate-change problem. Clearing lands for agriculture and allowing herds to overgraze will alter climates locally, whereas the emission of greenhouse gases can bring on global changes. People are not marginal members of the global climate system but central figures in it.

and if large sections of the forest are cut down, local rainfall will decrease.

Rain forests are also important to global climate. When trees are cut down, they no longer pull carbon from the air, and if their wood burns or decays, the stored carbon is released into the atmosphere. This action contributes to the human-induced global warming of the atmosphere.

There are other ways in which human activities affect the natural environment. Industrial activities put chemicals into the atmosphere,

CLIMATE VARIABILITY

Climate is weather averaged over time periods such as weeks, months, seasons, years, decades, centuries, or millennia. World and regional climates vary on all these time scales.

In tropical regions, temperature variations are not of major concern except in their highland areas, where frost occasionally can be a problem. Rainfall is of greater concern to farmers in the tropics because it determines the growing season and is vitally important for crop production and food security.

Rainfall is much more variable in arid and semiarid regions, and in such areas rainfall averages have relatively little meaning. For example, over a 35-year period in Gao, Mali, only two years had rainfall amounts near the long-term average during that period. A few years of heavy rainfall will certainly bring up the annual average, but overall the climate in such regions tends toward dryness.

Rainfall shortage in the form of extended drought can lead to food shortages and famine—a paramount concern for the world's developing countries. Drought by itself, however, seldom leads directly to famine. To generate a famine, drought frequently combines with societal problems such as high food prices, political corruption, or warfare, but drought usually takes most of the blame because it provides a handy scapegoat for political leaders whose countries face severe food shortages.

Outside the tropics, temperature is a primary determinant of the length of the growing season for different kinds of major crops. The length of the frost-free period is extremely important to farmers. For example, a late spring frost forces a delay in the planting of seeds, which in turn shortens the growing season and thereby increases the chance of crop damage if the fall's first frost comes early.

Another important facet of variability is an extreme event such as a drought, flood, freeze, or tropical storm—all too familiar to people in the United States and in other parts of the world. Extreme meteorological events are devastating to life and property, and reports about them are popular fare for the media, which lately tend to devote considerable space to climate-related news stories.

The phenomenon known as El Niño is an example of variability on a time scale of up to a few years. When El Niño events occur in the Pacific Ocean they often spawn weather disasters around the globe.

Volcanic eruptions also contribute randomly to climate variability for as much as a few years, depending on the size of the eruption. The gases and aerosols that eruptions inject high into the atmosphere can cool world temperatures for lengthy periods. Some of the largest and most devastating volcanic eruptions have become legends in climate history. These include Mount Tambora (1815) and Krakatau (1883) in Indonesia and Mount Pinatubo (1991) in the Philippines.

Mount Pinatubo, in perhaps the world's biggest eruption since the explosion of Krakatau, emitted about 25 tons of sulfur dioxide gas, creating aerosol clouds that kept ultraviolet rays from reaching the earth's surface and thereby cooling the global climate system for a few years. Although they are of relatively

In many parts of Asia monsoon rains enable farmers to grow rice. The summer monsoon sometimes brings good, soaking rains in June— sometimes not—and the variability in rainfall from one year to the next often takes the blame for crop failures. Grateful for the rain, farmers in Myanmar (below) hope to produce a rice crop large enough to feed their families and to sell for export to markets in Africa.

STEVE MCCURRY

short duration, such natural influences on variability tend to mask human effects on climate because they are so large.

On a time scale of decades, solar activity perturbs the earth's climate system. Of popular interest is the so-called sunspot cycle, which is approximately an 11-year cycle of minimum and maximum solar activity measured by counting sunspots. Many North American farmers believe it is associated with droughts that return to the Great Plains every 20 years or so, during times of minimum solar activity.

VARIABILITY / EL NIÑO

Following the 1982-83 El Niño event, newspapers ran numerous stories about "severe," "bizarre," "extreme," "crazy," and "anomalous" weather events disrupting human activities around the world and causing death and devastation on a local, national, and regional scale. Most of these events were linked (rightly or wrongly) to a phenomenon known as El Niño.

Folklore suggests that the warming of the Pacific's tropical waters was named El Niño for the Christ Child, because of a seasonal temperature rise that usually begins in December. Although the term originally referred to local conditions along the northwest coast of Peru, usage has been broadened to encompass all sea-surface warmings in the equatorial Pacific.

In the late 1950s, at the height of the Cold War, meteorologist Jerome Namias suggested that "people want to ascribe abnormal weather to almost anything.... Each generation seems to espouse some theory as to the changing climate. Of course, the present vogue [1958] is to blame things on nuclear tests." Paraphrasing Namias, one could say today that El Niño has become the focal point for our generation's search for an explanation of climatic variability.

Normally, off the coast of Peru cold deep water rises to the ocean's surface in a process called coastal upwelling. Strong southeasterly winds blowing along the shore and the earth's rotation push warm surface water away from the coast, allowing the deep, nutrient-rich cold water to well up and replace it. These regions are ideal breeding grounds for marine fish.

Sometimes, however, the trade winds in the western and central Pacific relax or reverse,

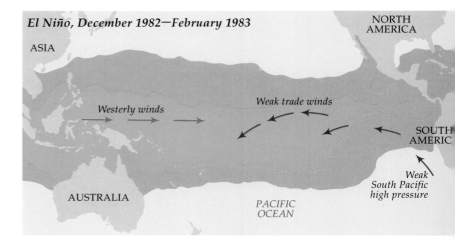

El Niño, December 1982—February 1983

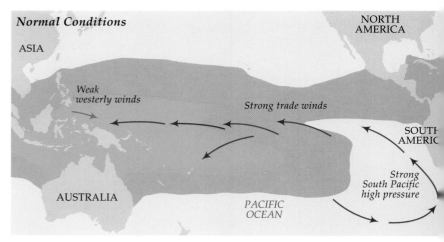

Normal Conditions

replacing the cold upwelled water with warm water in an event called El Niño. As the warm water invades the northern coastal regions of South America, it disrupts local weather patterns as well as marine biological processes, such as the food chain. Interruption of the food chain, in turn, adversely affects fish populations that thrive in the upwelling environment.

No two El Niño events or their effects have been exactly alike. Some events are short, with

During typical El Niño events, sea-surface temperatures rise across much of the equatorial Pacific (opposite, upper), as compared with non-El Niño years (lower). The phenomenon, occurring every four years or so, causes fish catches off Peru's coast to fluctuate. Anchovies, for example, decrease, but other fish seem to like El Niño's warmer waters. Near Pisco, Peruvians catch food fish (below) to sell in local markets.

LOREN MCINTYRE

minimal rise in sea-surface temperatures and with effects limited to South America's northwest coast. Accompanying them are local but often devastating effects, such as flooding from excessive rains falling on arid coasts or reduced fish catches in the waters off Peru and Ecuador.

When surface warming is prolonged in time (for two Southern Hemisphere winters and one summer), in space (surface temperatures become elevated over large expanses of tropical ocean), and in magnitude (warming can be 5 to 13°F higher than during non-El Niño years), the effects on atmospheric circulation and eventually on societies extend well beyond the region.

Thus, weather anomalies associated with El Niño are linked to various events, including droughts in Indonesia, eastern Australia, the West African Sahel, northeastern and southern Africa, northeastern Brazil, and parts of India; excessive rains in southern Brazil and on some

Pacific islands; fewer hurricanes in the North Atlantic; more typhoons in the Pacific.

Although people remain skeptical about scientists' ability to forecast El Niño a year in advance or to forecast its effects in every part of the globe, there is a silver lining to improved understanding of El Niño.

Once El Niño "locks in," it likely will run for 12 to 18 months, and some of its impacts occur after it is well under way. If scientists can

From 1991 to 1995, El Niño cut off the rain and shrank watering holes in parts of Queensland and New South Wales. Drought and bushfires plagued the land, devastating crops and livestock, and causing much economic hardship. Australians called this long dry spell the "Big Dry."

say with a high level of confidence that such an event has started, decision-makers can identify ways to prepare for probable impacts in their countries or economic sectors. Knowing several months in advance when El Niño will begin has

224

a potentially high value to leaders who use the information judiciously.

For example, the linkage of climate anomalies in Kenya, which grows and exports coffee, to El Niño events is not very clear. Kenyans, therefore, tend not to care about sea-surface temperatures halfway around the globe. But El Niño's effects on Kenya's coffee-growing competitors can reduce the ability of those other countries to meet the demands for their coffee in the international marketplace, thereby increasing Kenya's share of the market. Such information should be useful to Kenyan planners. Similar arguments can be made for corporations or other countries that import and export agricultural products.

Notable El Niño episodes occurred from 1972 to 1973, 1982 to 1983, 1986 to 1987, and 1991 to 1995. The 1972-73 event devastated Peru's fishing industry. The biggest El Niño in a century lasted from 1982 to 1983 and caught researchers by surprise when it began about September rather than in March or April. The 1986-87 event was the first one to be accurately and publicly forecast several months in advance.

The 1991-95 episode is controversial because a few researchers have suggested it was the longest one in the last two thousand years. Others disagree, noting that there have been El Niño events that have lasted more than a couple of years. Depending on how the information is analyzed, one could argue that there were three short events between 1991 and 1995 instead of one long one. Aside from this debate, Australia was plagued by the "Big Dry," a five-year drought that took a heavy toll on crops and livestock and resulted in numerous bushfires, one of which endangered the outskirts of Sydney, the federal capital and site of the Olympics in the year 2000.

Because of the simultaneous occurrence of El Niño and the massive eruption of Mount Pinatubo in 1991, the true adverse impacts of either event on regional weather conditions around the globe were difficult to identify.

El Niño (called the warm phase) is only a part of the cycle of changes in the sea's surface temperatures. The other part is called the cold phase or "La Niña." This occurs when surface temperatures become extremely cold. Another cycle under way at the same time in the equatorial Pacific is the seesaw-like pattern of pressure changes between Darwin (Australia) and Tahiti. This pattern is called the Southern Oscillation, which interacts with sea-surface temperatures to produce the ENSO (El Niño-Southern Oscillation) cycle.

So far, the lion's share of attention by researchers has been on El Niño events. Perhaps this is because the number of La Niña events in the past two decades has been relatively few compared with earlier times. For example, there was only one major La Niña between 1975 and 1990. Interest in La Niña may also have been low because such events are believed to be associated with weather and climate conditions considered to be "normal." Reports in the media reinforce this view with such statements as "El Niño is ending and the weather will return to normal." Yet La Niña events generally produce weather and climate conditions that are opposite those of El Niño. Viewing El Niño and La Niña as parts of the same phenomenon (ENSO) is therefore very important.

VARIABILITY / 1982-1983 EL NIÑO

El Niño from 1982 to 1983 was one of the most extreme episodes in a hundred years. Unusually high sea-surface temperatures in the eastern equatorial Pacific and many weather anomalies—droughts, floods, and untimely frosts—were associated with it.

Australia had its worst drought of the century. Agricultural and livestock losses, along with widespread bushfires in the southeastern part of the country, resulted in billions of dollars of lost revenues.

Severe drought plagued Indonesia, resulting in reduced agricultural output and in famine, malnutrition, disease, and death. The drought came at a time when Indonesians had been taking great strides toward self-sufficiency in food production. In the years immediately preceding El Niño, Indonesia had begun to emerge as a rice exporter. This drought, coupled with worldwide recession, huge foreign debt, and declining oil revenues, set back its economic development goals for the near term.

The United States endured devastating coastal storms and mudslides along California's coast, floods in southern states, and drought in the north-central states (reducing corn and soybean production). Salmon catches off the Pacific Northwest coast also declined sharply.

Large areas of Africa south of the Sahara were affected. Zimbabwe, usually a surplus producer and regional exporter of foodstuffs, suffered its worst drought this century. Food stocks were depleted; herds were slaughtered; and food assistance became necessary. The worst drought in two centuries forced the Republic of South Africa, an exporter of maize, to import 1.5 million tons of maize for animal feed in 1983.

For the second time in a decade, the West African Sahel was plagued by drought. While human and livestock deaths were fewer than those in 1972-73, the food situation was very poor.

Perhaps the worst drought-related impacts on food production took place in Ethiopia. At the time, the Marxist government was reluctant to admit to the global community the severity of its food problems. In fact, no government likes to admit that it cannot prevent famine. The combination of politics and drought can cause severe food problems in a country. And famine is what followed, with estimates of more than a million Ethiopians having starved to death.

El Niño also was blamed for droughts in Sri Lanka, India, Mexico, the Philippines, Brazil, and Hawaii; floods in Peru, Ecuador, Bolivia, and Brazil; and severe, unseasonal tropical storms in French Polynesia and Hawaii.

At local levels, El Niño was blamed for secondary effects ranging from major dust storms and bushfires in Australia to cases in the U.S. of spinal injuries in the West, encephalitis in the Northeast, rattlesnake bites in the Northwest, and bubonic plague in the Southwest. Each connection to a health problem had its own logic. For example, coastal storms altered the configuration of the seafloor off California. When surfers returned to favorite surfing spots they were injured in what had become shallower waters. In New Jersey, the wet spring created pools of water that are mosquito breeding grounds—hence, a greater number of mosquitoes and encephalitis cases. As for snakebites, Montana experienced unseasonably hot, dry weather in 1983. Field mice seeking insects entered populated areas and were followed by rattlesnakes.

Near Guayaquil, Ecuador, a washed-out bridge stops traffic along a major roadway (below). When Pizarro landed in Peru in 1531, perhaps El Niño's rains watered his troops as they made their way across the normally arid coast and went on to conquer the Inca. Today El Niño's heavy rains bring destruction to Ecuador and northern Peru, collapsing bridges, roadways, and buildings, and causing hundreds of deaths.

As a result, people confronted more snakes. In the Southwest, pleasant weather brought more tourists into wilderness areas where they frequently came into contact with prairie dogs and other rodents that carried plague-bearing fleas.

Before 1982, scientists thought they understood how El Niño develops. This event, however, did not follow the "typical" pattern. Each El Niño evolves differently, and this one showed researchers how much they needed to learn.

VARIABILITY / MONSOONS

One of the strongest expressions of the natural rhythm of seasons manifests in monsoons. The word has its origin in the Arabic word *mausim*, meaning "season." To the public, the arrival of the monsoon means the end of the dry season and the beginning of the rainy season. To meteorologists, however, winds are the key element in defining a monsoon. In southern Asia, southwest winds bring heavy rains from May through September. The winds reverse in winter, pouring cool, dry air onto Asia and dropping large amounts of rainfall on Indonesia, Australia, and other areas in the region with northeast coastlines. Agricultural societies here are heavily dependent on the timely appearance of the monsoon rains.

There are monsoons in Africa, South Asia, Southeast Asia, and China; and now scientists in the United States and Japan have identified monsoons in their countries as well. But, traditionally, monsoons are linked to the Indian subcontinent, where the summer monsoon is most important because it brings rain to more than a billion people in India and neighboring countries. This monsoon spans four months (June to September) and yields, on average, 80 percent of India's annual rainfall. The "normal" dates of onset and withdrawal of the monsoon vary in different parts of the country.

The monsoon rains are highly variable in time and space; rainfall amounts change greatly between years and between seasons. The total amount of rainfall over the whole of India varies by only 10 percent of average. In specific locations, however, such as the drier northwest part of the country, it can vary by 50 percent or more. India has a complete series of observations from a number of weather stations; some stations have records longer than a hundred years. The amount of rainfall, as well as its distribution in time and space, is important to Indian planners and farmers. In other words, the rain must fall in specific locations when the crops are in need of moisture for their growth and development.

The rains are crucial for the agricultural economies of India and its neighbors. These countries must feed large human and livestock populations and therefore depend heavily on rain. Some people say that the Indian national budget, in essence, is a bet on the monsoon.

Using 125 years of data, scientists can see that rainfall does not exhibit definite cycles. It does, however, exhibit periods of 30 to 40 years during which the monsoon alternates from long runs of good rains to long runs of poor rains. Every year since 1988, the monsoon has brought normal or above normal rainfall.

Prediction of the onset of the Indian monsoon motivated studies of sea-level pressure changes (and the development of the Southern Oscillation Index) in the Pacific Ocean. Henry Blanford, in the 19th century, tried to associate Himalayan snowcover with the monsoon. In the 1920s, Sir Gilbert Walker sought through statistics to identify the association of weather conditions throughout the world with monsoon rainfall. Today there is considerable interest in the possible connections between El Niño-Southern Oscillation (ENSO)—more popularly, El Niño—and poor monsoon rains over Asia.

Several Asian countries now use various statistical techniques to predict rainfall behavior, but to predict monsoon rainfall for small areas and for times within a season remains a challenge.

Rains of the summer monsoon must fall for farmers in India and neighboring countries to raise their crops and feed their families each year. These countries must also feed very large livestock populations. In Nepal (left), a planter's rice crop relies on abundant rainfall during a single growing season. When the expected rains don't come, however, the agriculture-based economies in this part of the world suffer greatly. Immortalized over the centuries by poets and other writers, the life-sustaining monsoon plays an essential role in many Asian cultures.

VARIABILITY / JET STREAM

The jet stream is an atmospheric phenomenon frequently talked about by television meteorologists but not generally understood by the public. The phenomenon is important because it often determines regional weather conditions around the world, and it is extremely useful to commercial airline traffic and military aviation.

This rapidly moving current of air is located about six miles or more above the earth's surface. A few miles thick and over one hundred miles wide, the air current encircles the globe. The jet stream is a product of the coming together of warm air from the equatorial region and cold air from the Poles, and it is strongest in the wintertime when the difference between temperatures at the Equator and the Poles is greatest.

Regionally, the air current straddles North America, influencing the formation of high- and low-pressure systems and guiding their paths across the continent as it flows from west to east. It has a major influence on local weather developments.

In winter, the jet stream drags cold Arctic air southward into the central and southern United States. In summer, it retreats northward to Canada, and the U.S. then enjoys much more temperate weather. The air current's springtime incursions into the central U.S. tend to bring severe storms, including tornadoes. At those times, cold, dry Arctic air and warm, moist Gulf Coast air meet, accompanied by strong jet-stream winds aloft.

Perhaps one of the worst North American winters in the 20th century, as well as one of the best examples of the enormous influence of the jet stream on the weather, was the winter of 1976-77. Abnormally warm coastal waters caused by shifting currents in the Pacific Ocean prevented the high-pressure ridge from making its normal migration southward. Anchored over the continent's western edge, the ridge diverted the jet stream northwest to Alaska, and with it the storms that would have watered the West. The jet stream then looped back to the south at half its usual speed, bearing the icy Arctic air that numbed the East. Drought in large areas of the western and southeastern United States, as well as harsh wintertime conditions across much of the country, resulted in major economic impacts. Shortages of natural gas led to the closing of factories and offices, putting 200,000 people out of their jobs and sharply increasing their fuel bills.

Many people, including most authorities, have claimed over the years that the jet stream was "discovered" by bomber pilots during World War II. Exceptionally strong headwinds reportedly slowed the pilots down when they were flying bombing missions over the Pacific Ocean. While this story seems on its way to becoming folklore, it has been challenged.

Some of the world's weather historians believe that the jet stream was, in fact, discovered in the early 1920s during an experiment with an unmanned balloon, which lifted off from Hampshire, England, and touched down four hours later in Leipzig, Germany. The calculated speed of this wind-carried balloon was about 120 miles an hour, and from that important piece of information, the researchers deduced the existence of a fast-moving stream of air at high altitudes.

Lined up by a hundred-mile-an-hour jet stream, high clouds cross the northern part of the Red Sea. In this view toward the south, Egypt appears at right, and the tip of the Sinai Peninsula fills the bottom left corner. Shifting jet streams steer weather systems along different paths on a daily basis. Longer-term shifts in these swift air currents can bring droughts, floods, and extreme temperatures to regions around the globe.

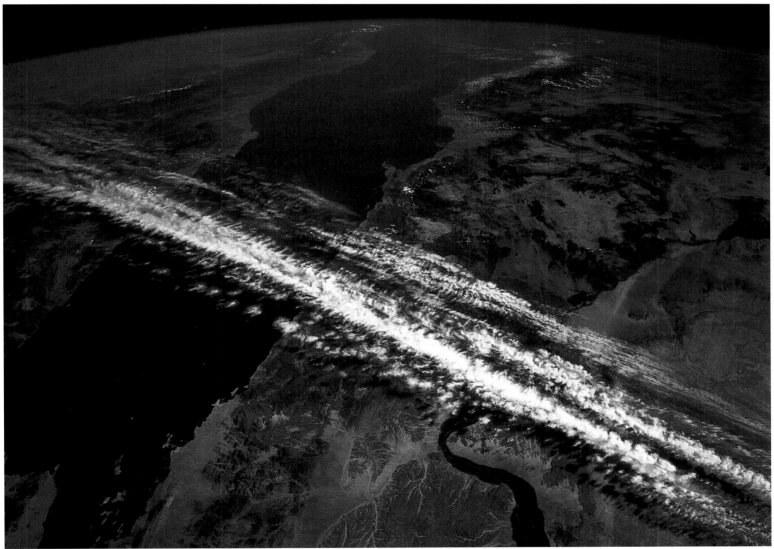

Today this current of air is used as a matter of course by pilots to shorten their travel time and to reduce their planes' fuel usage when making eastbound flights. They also use their knowledge of the jet stream to avoid its headwinds when flying westward, because those winds can significantly increase travel time. Clear-air turbulence at the edges of the jet stream, caused by the sharp contrast between winds inside and outside the rapidly moving current of air, is a feature that most pilots prefer to avoid as well.

VARIABILITY / FLOODS

Struggling in rising floodwaters, canoeists helping residents of Cedar Rapids, Iowa, suddenly need saving themselves. Many people thought that levees built during the past century would contain rampaging waters of the Mississippi River system. But in the summer of 1993, the worst flooding in a century destroyed those beliefs, forcing people in low-lying areas to abandon their homes and farms.

Floods have plagued humanity for millennia, from the biblical ordeal of Noah straight through to today. And in the United States, the mighty Mississippi River has been the setting for some of this country's more spectacular floods.

"In heat, mud and miasma, slaving men and sweating animals, toiling through the years, have thrown up 2,500 miles of huge, fortlike levees. Higher and higher they built them, hoping always that some day, somehow, they may achieve perfect flood control.… The levees, as built, have turned once vast, empty swamps into rich, thickly inhabited areas and added hugely to our national wealth…but when the river rises high enough it breaks them.… Now as I write this, it is breaking again. Today the most destructive flood in the annals of this rapacious river is rolling from Cairo [Illinois] to the sea." These are words written by Frederick Simpich while observing the Great Mississippi Flood of 1927. The estimated cost of property lost during that flood was one billion dollars (more than 10 billion in today's dollars). Some 800,000 people were forced to abandon their homes and were rescued from housetops, trees, levees, and railway embankments. Then Secretary of Commerce Herbert Hoover said of the flood, "It is the greatest peace-time disaster in our history." Similar statements have been made about the Midwest's Great Flood of 1993.

The Mississippi River basin drains about 40 percent (or 1.2 million square miles) of the lower 48 states, and parts of Canada as well. It is one of the most controlled rivers in the world, but it still floods.

In terms of the area affected, the 1993 flood

233

was one of the worst floods this century. Twelve midwestern states were directly affected by floodwaters; the toll included more than 50 deaths and severe damage to the environment and farmland. In addition to pesticides and fertilizers from farmland, toxic waste from former industrial sites and human waste from urban areas were flushed into the river.

The flood was surprising not only in its intensity and longevity, but also for having occurred in summer rather than weeks earlier—when snowmelt normally fills the Mississippi's tributaries. In 1993, the jet stream slipped south, bringing a mass of cool, dry Canadian air into contact with warm, moist air from the Gulf of Mexico. The resulting thunderstorms were then blocked by a high-pressure system stalled off the East Coast.

Flooding progressed relatively slowly from April through a peak in July, then picked up

During the Great Flood of 1993, water pours through broken levees near Valmeyer, Illinois (opposite). Farmlands went underwater, roadways washed out, livestock drowned, and homesteads flooded. Sandbags and water pumps protect a home from high waters in Ste. Genevieve, Missouri (below). To help save the town's historic district, volunteers came from around the country and joined residents in an all-out effort.

again in late August and September. Levees held for a long time, but more than 1,000 collapsed or were overtopped by rising waters. The high cost of the flood damages—between 15 and 21 billion dollars—came as a great surprise to local inhabitants and government agencies responsible for managing the river.

People here had not been thinking about floods for some time. In fact, a few years earlier some stretches of the Mississippi had dried up for several weeks. And it had been several years since the last major flood. When asked if his destroyed crops were insured, one farmer replied, "If you've farmed 70 years and never had this problem, you don't figure you need it. The levee is the crop insurance."

Many small towns and cities in the floodplain had been totally devastated. Once the waters had receded, inhabitants of Valmeyer, Illinois, a town about 35 miles southwest of St. Louis, Missouri, voted to relocate their town, with federal assistance, to higher ground.

Each town, county, and state has a saga to tell about coping with the floodwaters. The outpouring of moral and material support from other parts of the country and the daily coverage in the media made this a national disaster and not just a regional one. The Army Corps of Engineers, responsible for much of the engineering along the Mississippi, reported that "no other natural disaster in U.S. history affected or touched so many lives for so long a duration as the Great Flood of 1993. The resilience of the people in the region has been witnessed by how rapidly they put their lives and the lives of their communities back together, despite the severity of the disaster they had to rebound from."

VARIABILITY / DROUGHT

Drought is a climatic event familiar to millions of people. Just about every country, whether industrial or agricultural, rich or poor, has been plagued by it at one time or another. There are at least two types of drought: meteorological and agricultural.

Meteorological drought occurs when the amount of rain and snow in a region is less than expected for an extended time period. A sharp reduction in rainfall, however, does not always mean bad news for farmers, who still could have good harvests if the meager rains were to fall at the right time during crop growth. They also could withstand a drought if the market value of their surviving crops were to increase.

Agricultural drought takes place under two conditions: The amount of rainfall is either severely deficient, or it is adequate to produce crops but its timing is out of phase with a crop's growth-and-development moisture needs.

Cities are not immune from drought conditions, because they need water for municipal, domestic, and industrial purposes. While their sources vary, many cities depend on full reservoirs to meet their water needs for human consumption and power generation. In recent decades some cities have been forced to develop strategies to conserve water: Los Angeles in the early 1990s, San Francisco in the mid-1970s and New York City in the mid-1960s.

In North America, Europe, and Japan drought can have important economic impacts on farmers because of lost earnings and on consumers because they must pay higher prices for fewer agricultural products. While locally negative impacts can be quite severe, drought impacts at the national level may not be as bad.

The point is that droughts in industrialized countries are not life-threatening. In many parts of the developing world, however, drought is frequently a matter of life or death.

In sub-Saharan Africa, for example, many subsistence farmers grow barely enough food to feed their families. In a good year, they grow extra food they can sell in local markets or trade for needed goods, like cooking oil, pots and pans, and firewood. When drought comes to their region, they find it difficult to grow anything in their fields. If drought persists, they must migrate from their villages to search for food, work, or refugee camps where they are likely to find food donated by international organizations such as the Red Cross, the Red Crescent, Save the Children, Oxfam, the UN's World Food Program, and the U.S. Agency for International Development (AID).

But drought is not limited to Africa. It has also plagued northeast Brazil for centuries. Successive Brazilian governments have tried to develop policies to mitigate adverse effects on people and on the semiarid lands in the region. Dams, reservoirs, and irrigation have been the engineering fixes of choice. Government development agencies, including banks, were created to help break the poverty-drought connection.

Despite efforts to protect the northeast, drought continues to have negative effects in Brazil's poorest region. In search of a way to survive, people here migrate to the slums of major cities and into the rain forest, where they clear land to try to establish homesteads.

As we near the end of the 20th century, it seems that the best rain-fed agricultural lands are already being cultivated. If so, growing

In Mexico, a farmer surveys his parched fields. Without water, even the most fertile soils cannot produce vegetation. Because farmers who rely solely on rainfall to water their crops run the risk of damage by drought, many of them eventually turn to the practice of irrigation. Under severe and prolonged shortages of rainfall, however, irrigated crops also fail when water-bearing channels run dry.

ANNIE GRIFFITHS BELT

populations and increasing demands for food will put new pressures on uncultivated land. As population increases, people will be forced to move into areas less suitable for agricultural production under natural rainfall conditions. If history is a guide, they will move to these areas and plant the same crops using the same techniques they had used in better-watered areas. The risk of crop failure, however, will be increased by the mere fact that they are moving to drought-prone areas. Thus, we likely will hear more about drought-related problems as time goes on. This will not necessarily be the result of an increase in the frequency of naturally occurring drought or of global warming, but it will result from people trying to grow the same crops under marginal conditions.

Because drought occurs every year somewhere in the world, many efforts are being made to understand its causes and consequences. As a result, countries are shifting from being in constant need of international emergency aid to enacting appropriate policies to cope with drought and mitigate its impacts. As countries share their experiences and solutions, the burden of drought may lessen.

237

VARIABILITY / AFRICAN SAHEL

efore the early 1970s no one, except perhaps a handful of geographers, knew where the Sahel is located. This arid region is sandwiched between the Sahara's southern edge and a humid area south of the desert. It is an ecological zone that runs from west to east, cutting across what are now known as the Sahelian states: Senegal, Mali, Mauritania, Niger, Burkina Faso (formerly Upper Volta), and Chad. People in the region have been no strangers to drought. Before 1968, the Sahel had experienced at least two severe, prolonged droughts—one in the 1910s and the other in the early 1940s.

The 1950s and 1960s were wet decades compared with average conditions in the Sahel. Normally dry and brown, rangelands at the desert's edge turned green with vegetation. Noting the lush vegetative cover, government officials backed by their technical experts proposed that the areas closer to the Sahara, which had been used by pastoralists and nomads as grazing lands for their livestock, be cultivated as farmland. Farmers were enticed by colonial and, later, independent governments to extend their cultivated fields northward onto the rangelands, thus forcing the pastoralists to take their herds farther north toward the Sahara.

In 1968, a five-year drought began in the Sahel. With one crop failure after another, farmers had no choice but to abandon their fields and return to higher rainfall areas in the south. As for the pastoralists, they found themselves trapped with their dying herds on lands that had become part of the Saharan landscape. Researchers and policymakers alike suggested that the desert was moving south at rates of up to 30 miles a year. Photographs from the ground and images from satellites reinforced the belief that large stretches of the Sahel were being transformed into desert. At first, this transformation was believed to be caused primarily by climate change, but it soon became apparent that human activities were a leading factor.

In the midst of the drought, satellite images helped researchers identify a large pentagonal patch in a wide expanse of desertified land in Niger. Flyovers by aircraft and, later, an overland visit to the region confirmed it was rangeland that apparently had been well managed during the severe drought conditions; the land was covered by vegetation. The satellite images also showed large amounts of Saharan dust blowing at high altitudes over the west coast of Africa toward island countries in the Caribbean. The desert dust reduced the amount of sunlight that those countries received by about 10 percent.

STEVE McCURRY

Some 200,000 to 400,000 West Africans and about 12 million cattle perished. The plight of the people and the land eventually captured the attention of the international media, national policymakers, humanitarian organizations, and the public. In addition, photographs showed

LES STONE/SYGMA

Children suffer and starve during severe food shortages in sub-Saharan Africa. Survivors may make their way to cities or refugee camps, often becoming separated from the rest of their families. Lives disrupted, children such as this young boy (opposite) from Mali, West Africa, must cope with want at an early age. Ethiopian children (above) struggle to get by in a dry, denuded landscape. They await their daily rations in the refugee camp at Mekele in Tigray, Ethiopia. Many children and women in parts of drought-plagued Africa spend several hours each day searching for water and firewood.

large numbers of people on the move, abandoning their villages to search for food. News reports were filled with images of death and despair. Many people who survived the drought ended up in refugee camps, receiving only minimal levels of food, shelter, and medical care.

From 1968 to 1973 the food production capabilities in the West African Sahel were devastated by drought. Farmers tried to cultivate their lands against great odds, often having to plant seeds several times before enough rain would fall to allow the seeds to germinate. There was no assurance that after they were planted the crops would survive.

Nomads with their herds of cattle and camels were constantly searching for water and grasses. Technology had made it possible to dig deep wells so that water would be available year-round and not just in summertime. During this drought, however, nomads kept their herds around these permanent sources of water. Soon the animals had overgrazed all of the vegetation within a few days walk from the well sites. As a result, many of the livestock that died during the drought died from hunger, not thirst!

This drought in the Sahel caused the international community to develop a greater interest in finding ways to stop desertification processes, by which desert landscapes are created where none had existed in the recent past. As a result of the Earth Summit in Rio de Janeiro, Brazil, in June 1992, African governments sought and received renewed support for an international convention to combat desertification. Despite pronouncements of support at a UN conference two decades earlier, progress has been limited at best.

VARIABILITY / 1930s DUST BOWL

A Texas schoolboy in the 1930s witnessed an unforgettable sight: "I saw a woman who thought that the world was coming to an end. She dropped down on her knees in the middle of Main Street in Amarillo and prayed out loud: 'Dear Lord! Please give them another chance.'"

One 20th-century image that American minds surely will carry into the next century is of the Dust Bowl. Sand dunes piled up against farmhouses, barns, field equipment, and fences and formed in open fields. Fine particles of dust became airborne, and dark clouds blocked out the sun in midday. As the land turned to dust and blew away, multitudes of people gathered up everything they owned and headed West.

The combination of severe, repeated droughts and inappropriate agricultural practices in the Great Plains in the 1920s and '30s was responsible for the dust storms that plagued the American heartland. These storms came to symbolize change, sparking an exodus at first from the farms to the local cities and towns in the region, and later leading to a mass migration to California and elsewhere. These migrants, called "Okies," were not just refugees from Oklahoma; they also came from Texas, Kansas, Arkansas, and Missouri. Trekking westward to California on Route 66, the Okies stopped occasionally and worked as seasonal laborers to get money for gasoline and food.

The seeds for this environmental and economic disaster had been sown a few decades earlier. High prices for wheat and other crops

PHOTOGRAPHED AT NO MAN'S LAND MUSEUM, GOODWELL, OKLAHOMA, BY CHRIS JOHNS, NATIONAL GEOGRAPHIC PHOTOGRAPHER

Mountains of dust rise into the skies over Goodwell, Oklahoma, in June 1937 (left). When Americans recall the Midwest in the 1930s, they usually think of roiling dust storms threatening to smother whole towns. Such storms also menaced the region in the '50s, '70s, and '90s, but none topped the "Dirty Thirties" storms. Near Muskogee, Oklahoma, one family camps in a car (opposite). Drought and depression refugees, often called Okies, moved from place to place in search of work and food for their families.

during World War I and the years that followed prompted farmers in the Midwest to plant as much wheat as they could. They pulled up bushes, cut down trees, and plowed up grasslands so that they could use machinery to plant or harvest their crops.

They had removed the native vegetation that had been holding down the dry, sandy soil in the face of relentless winds. And when extended periods of severe drought visited the area, the winds strengthened and the topsoil began to blow away.

Jim Hodgson, President Nixon's Labor Secretary, was a teenager in Dawson, Minnesota, when the Dust Bowl days began. He recently recalled that "as the Rosebud Trio finished its bouncy tune, 'Happy Days Are Here Again,' the

announcer on WNAX in Yankton, South Dakota, turned serious. 'It's a happy tune all right, but we want all of you out there in Radioland to know something: This is the last time you'll hear that song sung on this station.' He explained: 'Not until it rains again, not until this midwest land of ours is green instead of brown, will there be anything for this region to be happy about.' Listeners sighed. For the third straight year their land lay parched and dry. Fate also found them caught in the third year of what became known as the Great Depression. But to the human spirit, the drought was clearly the more damaging of the two calamities. For denizens of the Dust Bowl, the banning of 'Happy Days' reflected vanishing hope."

Crop failures were widespread. An August 1934 issue of Newsweek at Home provided a bleak snapshot of the plight of farmers: "Corn was 80 to 90 percent without ears in Missouri, 75 to 90 percent beyond recovery in Kansas, 10 to 15 percent further damaged during the week in Nebraska, almost all destroyed in Oklahoma, South Dakota, and Arkansas."

Airborne dust made its way to Washington, D.C., where congressmen and Department of Agriculture bureaucrats began discussing how to respond to the continued threat of drought and the creation of desert-like landscapes.

Policies were proposed in the summer of 1934 and later put in place to protect the soils in the Great Plains. One policy was to construct shelter belts made up of thousands of miles of trees planted in rows. The trees and other shrubs would break up the hot winds, preventing them from drying out the soil as they passed overhead. The vegetation also would keep the

SO LONG, IT'S BEEN GOOD TO KNOW YUH

■ As part of an oral history project by the U.S. Library of Congress, singer Woody Guthrie was interviewed in 1940. He spoke about a dust storm on April 14, 1935, when people in an Oklahoma town huddled in their homes. Thinking that the end of the world had come, they began to hug and say their goodbyes. Guthrie said the inspiration for "So Long, It's Been Good to Know Yuh" was born while he sat in one of those rooms.

"I've sung this song, and I'll sing it again
Of the place where I lived, on the wild windy plain.
In a month called April, a county called Gray,
Here is what all of the people there say:
(Well, it's...)
Refrain: So long, it's been good to know yuh;
      So long, it's been good to know yuh;
      So long, it's been good to know yuh;
      But this dusty old dust is a-gettin' my home
      And I've gotta be driftin' along.
Well the dust storm came, it came like thunder.
It dusted us over, it dusted us under;
It blocked all the traffic and blocked out the sun,
And straightway for home all the people did run:
(Singin'...) *Refrain*
The sweethearts sat in the dark and they sparked,
They hugged and they kissed in that dusty old dark;
They sighed and they cried and they hugged and they kissed
But instead of marriage, they were talkin' like this:
(Honey,...) *Refrain*
The telephone rang. It jumped off the wall.
That was the preacher, a-makin' his call.
He said, "Kind friends, this may be the end,
You have your last chance at salvation from sin!"
Refrain
Well, the churches was jammed and the churches was packed,
But that dusty old dust storm it blew so black
That the preacher could not read a word of his text,
So he folded his specs, took up a collection,
(Sayin'...) *Refrain*."
—Woody Guthrie, ©1940

soils from blowing away. The American Tree Association called for the planting of 100 shelter belts 1,300 miles long across the Midwest from Canada to Texas. In support of the tree-belt concept, experts suggested that the winds of the plains, if uncurbed, might make a "Great American Sahara" out of rich farmlands.

The Dust Bowl Days, also called "the Dirty Thirties," coincided with the Great Depression. Writing for NATIONAL GEOGRAPHIC magazine 50 years later, William Howarth remembered that time: "Images of the Depression days still haunt our national memory: Banks failed, factories closed, workers lined up for free bread and soup. The Okies journeyed west with nothing, just hope and a pair of empty hands."

Sorting out the impact of a national economic depression from the regional impact of severe droughts is not easy. What is clear, however, is that the land in the Great Plains, in many instances, was being managed with little concern for the ecological limits imposed by the region's irregular rainfall and dry, sandy soils. In 1936, the U.S. Congress called for a review of the Dust Bowl situation in order to make policies to avoid a similar one in the future.

When the 1940s began, so too did the rains. People returned to the region to resume farming the land, and the Dust Bowl era began to fade into deep memory. The problems that had led to devastation in the 1930s had been resolved. Or had they?

In the early 1970s, when droughts plagued many countries around the globe, the American heartland remained unaffected. Grain prices soared, and farmers again began to plant everywhere they could. Many pulled out trees and shrubs in order to put more land into production. When drought hit in the middle of the 1970s, localized dust storms began to appear, and news stories reminisced about similarities to storms in the 1930s. In the spring of 1996, extended dry conditions in the southern Great Plains rekindled images of blowing dust and desertlike landscapes where usually productive agricultural lands were expected to be. Nevertheless, some people still believe that we will not have another Dust Bowl like that of the 1930s because, as one Oklahoman recently said, "Our farming practices are as close to perfect as you can get."

We must be ever vigilant about how we use our farmland. We must learn about and respect the ecological limits that vary from one location to the next. And those lessons must be passed on from one generation to the next, lest we forget the harsh lessons of the Dirty Thirties.

LAND CONSERVATION, AGAIN

■ A 1977 dust storm in California's San Joaquin Valley (below) recalls powerful storms of the past. When drought came to the Great Plains in the mid-1950s, dust storms billowed across the land and brought fears of a return to 1930s Dust Bowl conditions. While some people found it easy to point a finger at Mother Nature for the wind-driven devastation on the southern Great Plains, farming practices must share the blame. Land conservation practices put into place following the drought and desertification in the same region during the 1930s apparently had been forgotten. The return of a wet decade in the 1940s and the heavy demand for grains during World War II had provided strong economic reasons for many farmers to push their lands to the limit of production and well beyond. But in the 1950s, with the onset of yet another drought, farmers were forcefully reminded that they operate within a fragile environment. Once more, new soil conservation methods were put into practice.

SAM CHASE

VARIABILITY / SEVERE CLIMATE

Commenting on the damage caused in 1992 by Hurricane Andrew, the most expensive storm to strike the United States, Red Cross damage-assessment officer Nicholas Peake said, "If you put the 1989 Hurricane Hugo and the 1989 San Francisco Bay area earthquake together and doubled the magnitude of damage, that's what we have."

Hurricanes are tropical cyclones, like typhoons and cyclones in other parts of the world. Well known along the Gulf and East Coasts of the United States, the name comes from a West Indies word meaning "big wind." Hurricanes also bring storm surges, or seawater pushed onto the land many feet above normal. For example, when Hurricane Camille came ashore in 1969, it had maximum winds estimated at more than 200 miles an hour and a tidal surge of more than 25 feet. The combination of winds and surge can be devastating for many miles inland, especially if a hurricane strikes at high tide. In fact, gale-force winds can be felt hundreds of miles from a hurricane's eye.

The storms often move up the East Coast of the United States or enter the Gulf of Mexico. Hurricane Andrew, this century's third most intense Atlantic hurricane, destroyed lives and property in southern Florida in 1992, regained strength over the warmer waters of the Gulf of Mexico, and went inland near New Orleans.

In the 1960s, U.S. scientists tried to develop methods to reduce the intensity of hurricanes and to steer them away from land. If they could develop cloud-seeding methods to rob tropical storms of their winds, then lives and property would be protected. But not everyone wanted such storms diverted from their lands. Rains associated with hurricanes were important to farmers and reservoirs in Central America. Leaders there charged American scientists with robbing them of rainfall (some called it cloud rustling), and the research was called off.

Hurricanes, however, are not the only reason for the high level of destruction. Humans must share the blame for damages. People want to live in coastal areas known to be in the path of hurricane winds and tidal surges. Despite official building codes that call for construction that will withstand the fury of the winds, lax enforcement of codes can lead to substandard construction that, in turn, makes buildings and people more vulnerable to hurricanes.

And then there is the problem of how individuals respond to hurricane warnings. Some people prefer to ride out a storm in their homes, despite official concern for their safety. Others wait until the last minute to evacuate, hoping that a hurricane will change course or that a forecast is wrong.

On a larger scale, U.S. population migration in the last several decades has favored coastal areas. The population expansion has led to a boom in the development of infrastructure (houses, roads, businesses) all along the Gulf and the eastern seaboard, from the tip of Texas to Canada's maritime province of Nova Scotia, but particularly in the Southeast. Clearly, storms we've witnessed in the past hitting land in the same places would invariably cause considerably more damage to life and property. Roger A. Pielke, Jr., who researches weather impacts, reminds us that about the same number of people now live in South Florida's Dade and Broward counties as lived in all 109 coastal

Tropical storms target North America in this 1995 satellite image. That year's onslaught of hurricanes along the Atlantic and Gulf coasts of the United States set scientists to wondering whether it signaled global warming. Looking back, however, the past quarter century saw only a few hurricanes each year, while the years before 1960 witnessed many hurricanes. Perhaps the storms of '95 only harbinger a return to normal?

counties from Texas to Virginia in 1930. People have moved into harm's way, even though they continue to blame nature for the destruction. This fact is not a new revelation exposed in the wake of Andrew. NATIONAL GEOGRAPHIC writer Fred Ward commented on this distressing situation at the close of the 1980 hurricane season: "But developers, vacationers, and condominium owners continue to crowd the coast, and often refuse to compromise with the giant storms. Who is right may become clear only when the next killer hurricane finally comes."

The 1995 hurricane season, from June to November, was the second most active one in a hundred years of recorded history. The 1960s through the early 1990s have been relatively quiet with respect to the number of hurricanes that have occurred, and 1991 to 1994 was the quietest four-year period this century. Forecaster William Gray, a professor at Colorado State University, has suggested that we are entering an era of more numerous storms. His forecasts serve notice that when we develop coastal areas and replace old buildings with new ones, we should not forget hurricanes of the distant past. Only by reducing our complacency can we hope to reduce our vulnerability, increase our ability to cope with hurricanes and their aftermath, and live within the limits imposed by nature on our environmental setting.

CLIMATE CHANGE

Climate change can be considered objectively and subjectively. Objectively, it can be defined as variability, fluctuations, and change in the strictest sense of the word. Variability describes changes on a relatively short time scale, followed by a return to an average condition. Fluctuations represent nonpermanent changes that take place over decades. The strictest definition refers to longer-term changes over many decades or centuries, followed by a return to some expected average condition, or to a permanent change in average conditions prevailing for some period of time. Most policymakers today are concerned about this definition of climate change when they refer to global warming. The fear is that the climate regime is not just varying on some relatively short time scale but, in fact, is undergoing a permanent change.

Climate change can also be defined according to human perceptions. Humans tend to think in terms of shorter time scales, such as years and decades. Even if cooler or warmer years were to occur, people might view them as a permanent change in climate and not as variability or fluctuation.

Until the 1960s, an effective argument could have been made that studying climatology would be like watching grass grow: pretty dull and devoid of uncertainty or surprise. The field was viewed as a necessary but unexciting endeavor. In the early 1970s, fear of the resurgence of an ice age changed all that. After a few decades of relatively benign weather and climate, the period from about 1940 to the early 1970s was cooler than the preceding decades. The 1970s witnessed extreme adverse climate anomalies worldwide that, when linked with circumstantial and anecdotal information, suggested the earth was sliding toward the next ice age. For example, reputable scientists noted the following in support of a new ice age: The growing season in England was two weeks shorter than in previous decades; fish usually caught off the northern coast of Iceland were being found off its southern coast; hay production in Iceland had dropped by 25 percent because of cooler temperatures; armadillos that had migrated as far north as Kansas were moving southward, and so on. Food production dropped drastically in the early 1970s, as did fish catches. To show just how serious the ice-age fear was then, several books were published on the topic: *Fire or Ice, The Cooling,* and *The Weather Conspiracy.*

This view soon gave way to concern about global warming caused by human activities that heat up the atmosphere. For the past two decades scientists have been working to improve their understanding of the climate system so that they can identify long-term climate changes and the natural and human causes behind such changes. Just looking back at the past thousand years, climatologists have identified variability, fluctuations, and long-term trends. The warming that took place in the 11th and 12th centuries was a century-scale climate change. The little ice age from 1600 to about 1850 was a long-term trend toward cooling. The causes of these multicentury changes (trends) have not been identified with certainty. In the 20th century, there have been decades-long fluctuations: The 1900 to 1920 period was wet and cool; 1920 to 1940 was warm and wet; 1940

Fossil fuels—coal, oil, and natural gas—still power many of industry's manufacturing operations, such as this asphalt plant in Canada. Squarely implicated in the global warming debate, fossil-fuel burning releases carbon dioxide, which traps heat and warms the atmosphere, thus enhancing the naturally occurring greenhouse effect.

to 1970 or so was a cool period; from the mid-1970s the atmosphere has been heating up.

Today, policymakers in several countries worry about global warming. What will it mean for water resources and food security? What will it mean for economic development and dependence on fossil fuels? Some low-lying countries fear for their existence if sea levels rise.

Enough concern exists, despite the uncertainty surrounding the science of climate change, that world political leaders have signed on to the climate convention that was prepared for the Earth Summit in Rio de Janeiro, Brazil, in June 1992. The Scientific Assessments of the Intergovernmental Panel on Climate Change in 1990 and 1995 serve as the basis for the parties to the convention to hammer out agreements on coping with global warming through the management of energy, greenhouse gases, forests, and land use. Representatives meet each year at a Conference of the Parties.

Not all governments agree that human activities can alter the earth's climate, that fossil fuels are the culprit, or that they should sacrifice to resolve an environmental problem created by the "rich" countries who used up natural resources and jeopardized environmental quality to reach their high living standards. Many people think the rich countries that polluted the atmosphere with greenhouse gases should reduce their emission of such gases to ward off this global environmental problem.

CHANGE / NATURAL CAUSES

Several factors can bring about climate change. Scientists call them forcing functions because they make the climate system alter its behavior. They can be divided into internal and external factors: The former are inherent to the climate system; the latter are external to it.

External forcing functions include solar activity, variations of earth's orbit over very long time scales, and random volcanic activity. Global climate changes over the past thousand years have been attributed rightly or wrongly to one or a combination of these functions.

Many people believe that the number of sunspots (as a measure of solar activity) influences our climate. The sunspot cycle of 11 and 22 years is a prime example of a belief in this particular forcing function. Sunspots go through periods when they are numerous or relatively few, and people searching for some explanation of weather variations have resorted to linking the two phenomena. Using solar activity as an indicator, they try to forecast rainfall amounts in the agricultural heartland of the United States—the Great Plains—or estimate grain prices in the world marketplace as a result of drought that may be sunspot-related.

Although many scientists are skeptical of the linkage, people who look for cycles in nature believe that every 20 to 22 years or so drought returns to the Great Plains. They use the 1930s Dust Bowl as ground zero and step off 20-year intervals from there in both directions: Droughts plagued the midsection of the United States during the 1870s, 1890s, 1910s, 1930s, 1950s, 1970s, and most recently the mid-1990s. Today there is renewed interest in the scientific

community about the influence of solar activity on climate and weather.

In the 1920s, Serbian mathematician Milutin Milankovitch proposed the idea that changes in earth's orbital pattern have a major influence on the development of ice ages. The orbit places parts of the earth at varying distances from the sun in a given season, thereby causing different

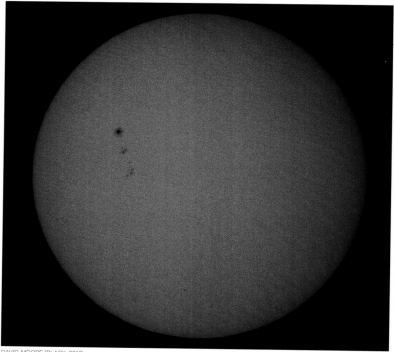

DAVID MOORE/BLACK STAR

climate regimes on geologic time scales of millennia and longer. Wallace Broecker of Columbia University succinctly stated Milankovitch's theory: "Changes in the tilt of the Earth's rotation axis and in the point in the annual cycle of the Earth's closest approach to the Sun alter the way solar radiation forces the climate."

NASA

Seen through a red filter on June 4, 1967, three dark spots, each larger than the earth, dot the surface of the sun (opposite). Many farmers believe sunspots somehow relate to droughts in the U.S. Great Plains and adjoining Canadian Prairies. While the 22-year sunspot cycle fascinates the public, it provokes controversy in scientific circles. In September 1994, astronauts aboard the space shuttle **Endeavor** *watched the eruption of Klyuchevskaya (left), a volcano on the Kamchatka Peninsula in Russia. Volcanoes spew ash and chemicals and may temporarily cool global temperatures.*

Ice cores are providing evidence of variability in the timing of the ice ages, and this variability—on scales of hundreds and thousands of years—cannot be explained on the basis of orbital theory.

For its part, volcanic activity affects climate on the order of only a few years. It can make identification of a climate change "fingerprint" more difficult because its influence on the world's climate can temporarily mask climate trends. A series of volcanic eruptions, however, could have a major influence on the global climate regime for periods much longer than just a few years.

CHANGE / HUMAN CAUSES

Human activities have been altering the natural environment for thousands of years. Since the onset of industrialization, however, the effects have increased dramatically, particularly on the chemical composition of the atmosphere. Societies everywhere have witnessed and perhaps contributed to changes caused by air pollution, acid rain, and ozone depletion. For example, the burning of fossil fuels (coal, oil, natural gas) releases large amounts of carbon dioxide (CO_2), a greenhouse gas, into the atmosphere; such gases trap long-wave radiation re-emitted by the earth's surface.

Carbon dioxide is not the only greenhouse gas produced by human activities. Methane (CH_4) is emitted by wetlands, by termites and other animals, and during the cultivation of rice. Nitrous oxide (N_2O), when used as a fertilizer in agriculture, also is released into the atmosphere. Chlorofluorocarbons (CFCs) were developed in the 1920s and have been used as foam-blowing agents, refrigerants, electronic component cleaners, and as ingredients in aerosol sprays. Although CFCs are also a greenhouse gas, their overall effect is still unclear.

As populations grow and economies industrialize, governments increase their energy consumption. Many of them rely more on fossil fuels in the absence of cleaner technologies. China, for example, has large coal reserves and plans to use them in its drive for economic development. Without other natural energy resources or the funds to import clean energy-related technologies, China must use its coal instead of expensive sources that do not produce carbon dioxide, such as solar or nuclear power. Other countries face a similar dilemma.

Many scenarios have been developed to identify the most effective and equitable way to reduce greenhouse gas emissions worldwide. But many people in Third World countries are convinced that the global warming problem has been caused by industrialized countries, and they believe those countries have the responsibility for solving it.

The industrialized countries have been pumping CO_2 into the atmosphere ever since the Industrial Revolution began in 1750. Their high levels of development and high living standards are the result of industrial processes. While good for society in many ways, the burning of fossil fuels and the production of other greenhouse gases have harmed local, regional, and global environments. Governments are now trying to cut CO_2 emissions without reducing standards of living, but their efforts face serious political difficulties.

Another contribution to the gas buildup comes from tropical deforestation in South America, Southeast Asia, and Central Africa. Deforestation contributes about 15 percent of the human-induced increases in greenhouse gases released into the atmosphere. Cutting down trees destroys vegetation that pulls carbon from the air and stores it, and burning them releases stored carbon. Industrialized countries are largely responsible for deforestation: Japan, for example, is the world's major importer of tropical hardwoods.

Many people in the developing countries rely on wood from trees and shrubs to cook their meals or to keep warm. In many locations the lack of firewood has led to what has been called "the other energy crisis." A political

Tropical deforestation (above) and wood burning contribute up to 20 percent of the human-induced greenhouse gases released into the atmosphere. Consumers in industrialized countries play a major role in deforestation. The gathering of firewood produces the "other energy crisis." Many people, such as women and girls in the African Sahel (below), spend several hours each day collecting wood to cook their meals, thereby denuding large areas of land around their villages.

leader from India once said, "Even if we have enough food in the future, how in the world will we cook it?"

Not all governments are convinced that human activities can really change the global climate. After all, the most pervasive greenhouse gas is water vapor. Carbon dioxide and other such gases are found only in small amounts in the air. How could these small amounts affect the global atmosphere?

Competing views on global warming can be divided into three categories. One group believes people are already changing the atmosphere and these changes are altering climate globally and locally. They believe extreme events, such as the droughts and floods in the 1980s and 1990s, are caused not by natural variability but by human-induced changes.

Another group does not believe humans

can alter the global climate by putting "puny" amounts of trace gases into the air. Noting that vegetation "likes" carbon dioxide (that's why we have greenhouses with enriched carbon dioxide environments), they also argue that agricultural production will increase with more CO_2 in the atmosphere.

A third group is made up of people who believe many uncertainties still remain in the science of climate change. They are impressed, however, by much of the circumstantial evidence linked to global warming, such as the fact that the hottest decade in more than a hundred years was the 1980s. Some years in the early 1990s joined the list of the top ten hottest years on record, but the trend was disrupted by cooling after the Mount Pinatubo eruption. While scientists have not answered some crucial questions (such as whether global warming will lead to more evaporation from the ocean and, in turn, to more clouds and cooling of the global climate), this group believes "no-regret" precautionary actions could be taken. Actions such as conserving energy, using alternative energy sources, and reducing fertilizer use would be beneficial whether warming is occurring or not.

While most policymakers show an interest in global environmental change, they are primarily concerned about climate changes that affect their political jurisdictions. In several locations around the world, human activities have been altering climate locally and regionally, and perhaps one of the most obvious examples is the so-called urban-heat-island effect.

The construction of cities and the concentration of energy usage in them have generated unnatural climates in urban settings. Some researchers have found that a heat island produces more precipitation within a city and downwind from it. St. Louis was used to prove this hypothesis. The heat-island effect has meant that temperature recordings have to be adjusted for changes in records that followed the warming of the cities. Because of the effect, some cities are showing a 3°F warming.

Land-use changes—deforestation, clearing, cultivation—can also alter local and regional climates. According to Russian scientists, certain farming practices increase rainfall. They believe that wide-scale irrigation can produce enough evaporation to favorably alter the local climate, making the air more humid.

Tropical deforestation, whether for commercial timber harvesting or for homesteading, not only contributes to global warming, but also can change the local climate. For example, about half the rainfall in the Amazon basin results from evaporation from the forest's vegetation. Cutting down that vegetation reduces evaporation and can ultimately reduce local rainfall amounts. In fact, human activities can create arid landscapes in tropical rain forests because, once the vegetative cover has been removed, the soils are too fragile to withstand the intensity of raindrops, the sun's rays, and winds. In addition, rain tends to rob soils of their nutrients. The result is a barren landscape exposed to the raw elements of nature.

Food production efforts can also lead to the destruction of local and regional environments, which can then alter the climate. Cultivating areas that are only marginally suitable for sustained cultivation eventually impoverishes the soil and modifies climate at the local and

Coal mining (foreground) and a coal-fueled power plant (background) near Rock Springs, Wyoming, release gases and particulates into the atmosphere. As a result, the quality of local and regional air declines, and acid rain may occasionally fall from the skies. The effects of industrial emissions can reach far beyond regions—quite possibly altering atmospheric temperatures around the globe.

PAUL CHESLEY

CLIMATE MODIFICATION

■ Throughout the 20th century, many ideas for changing the climate have been proposed to government leaders. Such schemes have included the damming of the Bering Sea, deflecting major ocean currents, towing icebergs, flooding desert depressions to create inland seas, diverting rivers, large-scale irrigating of arid lands, and blacktopping arid coastal areas. Each scheme has a logic that, in theory, sounds as if it might work. Interestingly, while we have been trying to devise ways to improve regional climates for human benefit, we have been destroying land known to be productive. The difference between attempts to alter regional climate and the global warming that may be occurring because of greenhouse gas emissions is that the former is planned and the latter is an uncontrolled experiment carried out by civilizations, the outcome of which remains uncertain.

regional levels. And growing the wrong crops in places that are wholly unsuitable for them can lead to an increase in the frequency and intensity of droughts.

The same situation arises on rangelands. Too many cattle, camels, or goats can destroy the vegetative cover. Population and political pressures push more people into these areas, and when they use the land, often in ignorance of local constraints imposed by natural conditions, they set themselves up for failure. These situations are analogous to events that could occur on a larger scale if global warming occurs.

253

CHANGE / DEFORESTATION

Tropical forests are disappearing at alarming rates. The media report deforestation in terms that people can understand: An area the size of a football field is destroyed each minute in the Amazon; every week 113,000 soccer fields are cleared; an area the size of Belgium, the Netherlands, and Luxembourg is deforested globally each year. Such statements convey a sense of urgency, but they do not provide a sense of accuracy.

A plethora of photos now show denuded landscapes in the Northern and Southern Hemispheres. While the media have focused on deforestation in the Amazon basin, other locations are at equal or greater risk. The lowland rain forests of West Africa have been decimated. Haiti is virtually deforested. The uplands of Thailand have been stripped. Madagascar's

While Brazil's Amazon receives the lion's share of attention when it comes to destruction of tropical rain forests, other regions also reel from deforestation. In eastern Borneo (below), commercial logging of tropical hardwoods meets the needs of buyers in industrialized countries. As regions lose their favored trees, pressures increase on remaining forests.

forests are being rapidly destroyed. The forested uplands of India's major river basins have also been destroyed. It is noteworthy to point out that Brazil's high rates of deforestation during the 1980s were mostly based on spectacular changes in the state of Rondônia. The BR 364 highway, financed by the World Bank, was constructed there without environmental impact assessments. As a result, it served as a conduit for urban migrants wanting to improve their quality of life by moving into the new territory.

Latin America contains 57 percent of the world's tropical rain forests. Asia claims 25 percent, and Africa, 18 percent. No one doubts that the forested area, estimated to occupy about 13 percent of the earth's land surface, is on the decline. The reasons, however, vary from region to region. In Latin America, rain forests are being converted to pastureland and farms; in sub-Saharan Africa, they are being cleared for timber export and to meet demands for farmland and firewood; in Southeast Asia, forests provide the hardwood products exported to industrialized countries. Other societal factors include a lack of equitable land reform programs, chronic poverty, poor forest management practices, government corruption, and large government development projects such as dam construction.

Although the situation may have been different in each case, the consequences have been similar: flooding downstream, high levels of soil erosion and sediment in rivers, soils robbed of fertility, and barren, degraded landscapes. Downstream flooding and subsequent death and destruction usually ensue even in average rainfall years. The main causes are commercial logging and land-clearing for agricultural settlement and cattle grazing. An ambitious plan for resettlement has been proposed by Indonesia. Its transmigration policy is designed to relieve overcrowding and poverty in Java by moving millions of people into virgin forests in western New Guinea (called Irian Jaya).

Rain forests are an integral part of the global climate system. They act as a major carbon sink, pulling carbon out of the atmosphere during photosynthesis and storing it. When cut trees are burned or left to decay, carbon is released into the atmosphere as carbon dioxide, a major greenhouse gas. Scientists estimate that up to 15 percent of the human contribution to atmospheric carbon dioxide comes from the effects of worldwide deforestation.

Governments in the tropics, like others around the world, are developing strategies to improve the standards of living for their citizens. Often these strategies conflict with the desires of some members of the international community eager to preserve the natural environment. Rain forests are of special concern to people worried about a loss of biodiversity.

Pharmaceutical companies see potential in the commercial exploitation of biodiversity for medical applications. The rosy periwinkle in Madagascar, for example, has been used in a cure for childhood leukemia. Pyrethria, from sub-Saharan Africa, has been used in the manufacture of insecticides. Yet, to date the developing countries have derived little financial gain from such discoveries, so their incentive to protect the forests is minimal. Perhaps a way to get them to conserve forests is to provide economic incentives benefiting their societies generally.

CHANGE / DESERTIFICATION

At the Earth Summit in Rio de Janeiro, in June 1992, African nations called on the UN General Assembly to prepare a Convention to Combat Desertification in Those Countries Experiencing Serious Drought and/or Desertification, Particularly in Africa. A convention was drawn up and adopted in Paris on June 17, 1994, and it came into force at the end of December 1996. It says that "desertification means land degradation in arid, semi-arid and dry sub-humid areas resulting from various factors, including climatic variations and human activities." That statement is today's official definition of the term.

Desertification is the creation of desertlike conditions where none existed in the recent past. By undermining the land's productivity, it contributes to poverty. Desertification is closely associated with arid lands along desert fringes, but it also occurs in high rainfall regions such as South America's Amazon basin. It might be called a mega-concept that encompasses several degradation processes, including wind and water erosion of the soils, salinization, overgrazing, waterlogging, and excessive wood cutting.

The causes can be natural, human, or a combination of both. Regional climates go through periods of fluctuation with wet or dry periods that are decades or centuries long. In the ancient past, desertification processes depended on interplay between climate and the land. In the past few thousand years, however, the interplay has expanded to include human activities. During ancient times soil impoverishment took place over centuries or millennia, but today's degraded landscapes seem to have been created in a few years or decades. Humans have made the difference with their technology and land-use practices.

Countries are now quite active in combating desertification. In North America, desertification is primarily an economic problem, and the governments there can use financial and technological resources to cope with its causes and effects. A different situation exists in the Third World, where the loss of agricultural land to encroaching desert can mean starvation.

Many people in arid and semiarid areas live from one rainy season to the next. After they pay their debts, they have little left to feed their families until the next year's harvest. When soil fertility declines, food production declines, and by the next growing season the people often are too malnourished to work their fields efficiently. Again production declines. Eventually farmers look elsewhere for productive soils. But new lands with good soils are not likely to be available, so they move onto marginal lands that are easily degraded after only a few years. Ultimately, they may abandon the land and seek work on commercial farms or in urban areas.

Decades ago, when population densities were lower, farmers could migrate to new areas and let their lands lie fallow. After several years, the fields would recover their fertility, and the farmers would return. Today, with virtually all the rain-dependent arable land in production, they must farm their plots on a continual basis.

Poor countries are most vulnerable to desertification because they do not have the funds to combat it effectively. They need training programs and technologies to arrest these processes. Teaching people to avert desertification is cheaper than reclaiming land already degraded.

Herders lead their skinny cattle in West Africa's Sahel, a common scene in an arid land with erratic annual rainfall. A few decades ago, newspapers ran headlines about the Sahara's southward march at 20 to 30 miles a year, overtaking rangelands and farms. Overgrazing practices and poor cultivation methods, as well as a scarcity of rainfall, caused destruction of the land and desertlike conditions.

ALAIN NOGUES/SYGMA

Desertification is a long-term, low-grade, cumulative problem. Negative changes that occur each day are virtually invisible and do not require immediate attention from policymakers. After several years, however, degradation is quite apparent. By then the situation is a crisis, and large sums must be spent to reclaim the land or it must be abandoned altogether.

Desertification's adverse effects are often hidden during rainy periods, but poor land-use practices are highlighted during dry periods. While drought is frequently blamed for the loss of soil fertility, it often only worsens an already bad situation. If climate change increases the number and intensity of droughts, desertification processes will likely accelerate and become more widespread. It is important, therefore, to better understand these processes and to identify ways to avoid them in order to minimize the effects of future droughts.

CHANGE / ARAL SEA

The Aral Sea is drying up. Until 1960, its surface area was about 26,000 square miles, its volume about 265 cubic miles, and its depth about 175 feet. The sea's area now is about 14,000 square miles, its volume about 80 cubic miles, and its depth about 120 feet.

The sea basin lies deep in the heart of the Eurasian landmass. It was within the Soviet Union until 1991 when the political entity was dissolved. Now parts of it are in five newly independent republics: Uzbekistan, Kazakstan, Tajikistan, Kyrgyzstan, and Turkmenistan.

The climate here is continental, with hot, dry summers and cold winters. Over the sea, evaporation rates are quite high, especially in the summer months. Seventy percent of the yearly rainfall comes during winter and spring, and the highest amounts fall in the foothills. Embedded in an arid zone, the sea influences local climate several miles from its shores.

Two major rivers feed the Aral sea—the Amu Dar'ya and the Syr Dar'ya. Although called a sea, the Aral is really a terminal lake, whose origins are in some of the world's highest mountains: the Pamirs and Hindu Kush to the southeast; the Tian Shan and Altay in the southeast and east. The sea is also bordered on the south by two deserts—the Kara Kum and Kyzyl Kum.

The rivers' flow for much of the sea's history had been unaffected by human activities. For several centuries, water in limited amounts had been diverted to irrigate the dry but fertile desert soils. Diversions slowly increased until 1960, when political decisions were made by Soviet leaders to sharply increase the amount of desert lands devoted to cotton cultivation.

GERD LUDWIG

Larger diversions would be required. In addition, the Soviets decided in the mid-1950s to construct a record-setting canal across the Kara Kum, drawing water from the Amu Dar'ya.

The new republics depend on Aral water for agricultural production: In Uzbekistan cotton and rice prevail; Kazakstan grows cotton and rice, and raises cattle; cotton, hydropower, and processing industries are important to Tajikistan; in Turkmenistan cotton and Astrakhan sheep are important; Kyrgyzstan focuses on fodder production and cattle.

From the late 1970s till the early 1990s, no water reached the Aral from the Syr Dar'ya,

Fishing boats lie where they landed when the Aral Sea shrank from these shores. Over a 30-year period, leaders in the former Soviet Union encouraged the diverting of water from the sea's two major feeder rivers and used it to irrigate desert sands for cotton and rice production. The likely result? By the early 21st century, the Aral may be a sea without water—the "quiet Chernobyl," according to some scientists.

and no Amu Dar'ya water reached it in the late 1980s. The Aral Sea is dying. With small amounts of river water reaching the sea, there is not enough inflow to balance losses from evaporation. As a result, its level continues to drop.

Debate over the Aral Sea has raged throughout the 20th century. To some observers, letting the water go unused for economic development was simply a waste of water. It was evaporating; why let it go to waste? Few people objected to its use, and no one sought to preserve it because of a perceived intrinsic value, such as just being a sea in the midst of a desert, a sea supporting various kinds of flora and fauna.

Environmental impact statements were not part of the Soviet way of thinking in those days. In fact, government actions regarding natural resources were in essence based on all-out exploitation of them. Society was responsible for taking these resources from nature whenever the needs of society demanded it.

Industrial and other municipal wastes have also found their way into the rivers of the Aral basin. Groundwater that is used extensively for municipal, industrial, and drinking purposes is heavily contaminated throughout the region, and this sad fact is largely responsible for the poor state of health in the area.

CHANGE / SEA-LEVEL RISE

A major concern about global warming is its effect on the level of the oceans. All projections by scientists note that if the atmosphere heats up, the level will rise. Glaciers will melt and seawater will expand because of the heat it will receive from the warmer atmosphere. Estimates of a sea-level rise range from a few inches to three feet by the end of the 21st century. Increases of this magnitude can have major impacts. For example, a rise of one foot would cause most beaches to erode by 100 to 200 feet.

In the 20th century, the annual increase in sea level has been less than a tenth of an inch. A rise of two feet would inundate more than 80,000 square miles of wetlands and lowlands in

MARIA STENZEL

WEST ANTARCTIC ICE SHEET

■ If polar sea ice were to melt, the level of the world's oceans would not increase. The effect would be similar to an ice cube in a glass of water: When the ice cube melts, the water level in the glass does not change. The picture is much different, however, for the West Antarctic ice sheet, which contains an estimated 770,000 cubic miles of ice (by comparison, the Greenland ice sheet contains about 720,000 cubic miles). The West Antarctic

ice sheet is anchored on the ocean floor, but it has ice ledges that are more than 1,500 feet thick, with a large portion jutting out into the ocean. If the ice sheet begins to melt as a result of warmer ocean temperatures, a large part of it could become unstable and surge into the sea, raising the level of the ocean by about 20 feet and inundating many of the world's coastal areas, including deltas, estuaries, and several major cities. Scientists have debated the possibility of the disintegration of the West Antarctic ice sheet since the late 1970s. The prevailing

view today is that the ice sheet's collapse is not likely to happen in the next few centuries, and even after all that time, the chance of a collapse is still quite low. However small, the risk is there nonetheless. Scientists use computer models to glimpse what might happen to the polar regions during global warming. Research and historical assessments indicate that a 1°F warming in the U.S. could be accompanied by a warming of 3° to 4°F at the Poles. Otherwise, the specific effects of global warming on the polar regions are still not clear.

the United States. Many cities situated along coasts or rivers would be affected. River deltas are highly productive regions for agriculture

Talk of global warming leads to fear of rising sea levels, especially in low-lying areas. In the Mississippi Delta (below), people must also deal with rising water caused by sinking land: Subsidence results from withdrawal of groundwater and compaction of sediments.

and marine life, and they would suffer greatly. Small island states are deeply concerned about sea-level increases that will make them more vulnerable to inundation by tropical storms.

A World Meteorological Organization report noted that there will be "an increase in flooding, in the upward movement of the water table, and changes in the pattern of coastal deposition of sediments and of coastal vegetation. More soil erosion may occur and water quality may be impaired. Perhaps there will be abandonment and migration from the coastal zones."

Countries such as Bangladesh will suffer as much from inundation by sea-level rise as by storm surges. Much of the country is barely above sea level and seasonal flooding is not unusual. If the sea rises, a large portion of the country will be permanently underwater.

The Netherlands, having reclaimed much of its land from the sea, has been living for centuries with the threat of flooding. The country says it has the technology to withstand sizable increases in sea level and that it can help other low-lying areas protect themselves. Dutch engineers and planners are sharing their knowledge and technology in such distant places as the Maldive Islands in the Indian Ocean and the Mississippi Delta in the Gulf of Mexico.

In the U.S., the sea level is already rising in some Louisiana parishes, not because of global warming but because the areas are subsiding. The subsidence results from the withdrawal of groundwater needed during oil exploration processes. How Louisiana responds might provide insights into how other low-lying regions might deal with sea-level changes brought on by global warming.

CHANGE / CASPIAN SEA

The Caspian Sea is the largest inland body of water. The Volga River, which drains about 40 percent of the European part of Russia, flows into the Caspian and is its major source of water. Until 1991 it was a sea shared by the Soviet Union and Iran.

Following the breakup of the Soviet Union, three new countries were formed along the Caspian Sea: Kazakstan, Turkmenistan, and Azerbaijan. The change is an important one—the resources of the sea now must be managed by five countries instead of two. This is not an easy task, because each country has its own economic development goals.

The resources in the Caspian include oil reserves and caviar-producing sturgeon populations. The Caspian sturgeon provides about 90 percent of the world's caviar, and the sea's oil fields are believed by some experts to be even richer than some sites in the Persian Gulf.

Municipal, industrial, and agricultural wastes are carried to the sea from the various rivers flowing into it. Pollution also directly enters the sea from oil spills and discharges from offshore rigs. These pollutants have a negative impact on the environment, including the sturgeon population.

The level of the Caspian Sea dropped about six feet from the 1930s to the 1970s, and people moved onto the newly exposed seabed and built houses, roads, farms, and so on. For a few decades some experts feared that the Caspian would eventually disappear. But then, in 1978, the sea level rose abruptly, and the waters have continued rising ever since. This slow but steady increase reached eight feet by the middle of the 1990s, and many Russian scientists believe that the rise will continue until 2010.

Over the past 20 years the sea has flooded low-lying coastal areas in every country bordering it. Farms have been inundated; buildings have been destroyed; beaches have been eroded; roadways and railroads have been submerged; and people have been evacuated. Strong winds blowing onshore cause storm surges, driving the elevated seawater farther inland than normal and causing greater devastation. As the sea rises, so too will the damage to human settlements.

Considerable speculation has been raised about why the Caspian Sea has been on the rise. Some experts have said that higher amounts of rain and snowfall in the Volga River basin are the cause. Others have suggested that perhaps it is the result of tectonic plate movement at the earth's surface. Such movements would cause the seabed to rise, giving the appearance of a rise in sea level. Still others have suggested that the diversion of waters from the Aral Sea basin to irrigate the surrounding Central Asian deserts has led to seepage into the groundwater, which has

GEORGE F. MOBLEY

Moonlight reveals high-density oil rigs near Baku, Azerbaijan, one of five countries along the Caspian seacoast. Because the sea's oil reserves may rival those of the oil-rich Persian Gulf, many international oil companies court the Caspian countries for exploration rights. These waters also present a threat though. In the past 20 years, the sea level rose 8 feet, inundating oil facilities, roads, and buildings.

then flowed into the Caspian. This is a simple explanation for two environmental changes in the region: the rise of the Caspian Sea and the simultaneous decline of the Aral Sea level.

Aside from providing a realistic picture of the impacts of sea-level rise on society, the Caspian situation can also serve as an example of the possible impacts on low-lying areas of sea-level rise caused by global warming. In other words, watch what happens to ecosystems and society here for lessons on what could happen if warming raises sea levels worldwide.

CLIMATE MONITORING

Evidence seems to show that from the earliest days of their existence humans have been interested in understanding the climate. The climate has a major influence on the foods people eat, the places where they live, the clothes they wear, and how they interact with each other. For these reasons and because of basic human curiosity, people have been gathering information about weather and climate for ages. In fact, much of the information we have on the climate system has been handed down to us over the centuries.

Perhaps one of the earliest recorded treatises on climate was *Meteorologica,* written by the Greek philosopher Aristotle around 350 B.C. While many of his observations and hypotheses were proved incorrect later, his work was held in high regard until the Middle Ages. In the years since that time, much more reliable knowledge has been slowly gathered about the workings of the climate system.

Today most of the world's governments are interested in learning about climate variability and change primarily because they believe research will generate usable information for policy-making. With this hope in mind, governments try to provide at least a few billion dollars each year for researching and observing the climate system. Leaders want to know as much as possible about the future climate, and scientists try to obtain that information for them in several ways: by using computer models; making observations of the land, sea, and space; studying paleoclimatic evidence; and reviewing historical records and analogies.

Much of the information is gathered not only for the purpose of understanding the global climate system, but also for learning how to forecast its behavior and the behavior of its parts on different time and space scales. We now have a large arsenal of technologies and techniques for researching, monitoring, and forecasting the many aspects of variability and change.

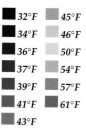

General Circulation Models

Researchers use sophisticated computer models in their attempts to understand the workings of the climate system and the interrelationships of its components. Such models, which are called general circulation models (GCMs), are used by oceanographers, atmospheric scientists, and other experts in the field of climatology. In recent years much emphasis has been placed on linking models of the ocean and atmosphere to get a more accurate glimpse of the air-sea interactions affecting global climate.

One of the most important characteristics of GCMs is that they can process large amounts of data. This aspect makes them a highly preferred tool for trying to understand changes in atmospheric processes. Because they can be used to simulate past climate variations with a certain degree of reliability, they are also used to make general, long-range projections about the world's climate. For example, by varying conditions in the model, such as the amounts of different combinations of greenhouse gases in the atmosphere, global temperature changes can be projected with some degree of confidence.

Perhaps the greatest problem with GCMs is that researchers have to assume that there are average conditions over large geographic areas called grids. Average conditions, however, do not reflect reality—considerable variability can

In their efforts to grasp the workings of the climate system, researchers turn to general circulation models (GCMs) run on computers that can quickly process large amounts of climate-related data. While GCMs may not yet be reliable in projecting regional climate changes, they have better success projecting temperature change at the global level. The figures at left compare experimental results from GCMs prepared in 1984 and 1989 by the National Center for Atmospheric Research (NCAR) in Boulder, Colorado. Containing a more realistic ocean component, the 1989 model (lower figure) shows associated air-sea interactions that drive the global climate. According to these climate-change simulations, the greatest amount of warming occurs in the high latitudes, or polar regions, during winter in the Northern Hemisphere.

occur within a large area. Hence, there is considerable uncertainty in GCM-generated projections about regional temperature changes over decades or centuries.

While general circulation models may be reliable for projecting changes in global climate, they are much less so at the regional level. Most policymakers are interested first and foremost in what happens in their regions and especially about how their constituents will be affected. Researchers' attempts to develop regional climate models that cover smaller areas more accurately are only in the experimental stages. These regional models (also called limited area models) include a higher density of information

for a particular area, and they can be inserted into a larger GCM to provide better information about changes in the global atmosphere.

General circulation models contain much of the science that researchers know about the atmosphere, the oceans, or their interactions. By using the GCM tool, scientists can learn more about the climate system than by using any other tool. For example, the historical record of climate and the oceans is relatively short. As a result, societies have not yet experienced a full range of variability that might occur in the global climate system. Using the GCMs, modelers can explore a broader range of possible climate variations to which societies might be subjected.

MONITORING / OBSERVATIONS

Because models run on imperfect theories about climate, they require lots of information about the climate system and how its components interact. For that reason, computer modelers are major consumers of climate-related observations.

Major efforts are being made to collect information just about everywhere on air temperature, rainfall, wind, pressure, sea-surface temperature, and other meteorological and oceanic aspects. Some information is collected on a routine basis, while other information is gathered during short field experiments involving many countries. Scientific experiments are often too costly for one agency or one country to carry out alone. Besides, the atmosphere doesn't stop at an international border.

Every day people take in information about their local climate, but they do so in a haphazard way. Without written records, their knowledge becomes blurry and distorted by their perceptions. Their memories can do funny things when it comes to recalling climatic events. So, this winter becomes "the worst I have ever experienced," or that hurricane was "the strongest ever," or the frost "came too early this year." But are their recollections reliable about the magnitude or impact of climate?

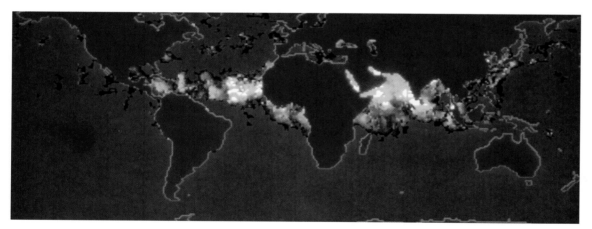

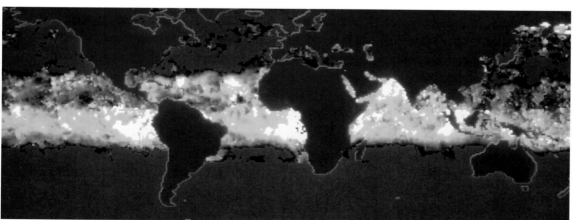

Satellites provide global coverage of our planet's atmosphere, oceans, and land. In 1991, they helped scientists monitor the eruption of the Philippines's Mount Pinatubo, which ejected about two cubic miles of ash particles and 20 million tons of sulfur dioxide into the atmosphere. Satellite data represented at left show an increase in sulfuric aerosols primarily over the Indian Ocean (upper). Two months later, the shroud of aerosols encircles the globe (lower). The sulfuric aerosols reflected sunlight back into space, lowering global surface temperatures by about one degree Fahrenheit. From key locations around the world, scientists frequently launch balloons carrying devices that measure changes in the atmosphere's chemical constituents (opposite).

Professionals carefully observe the climate system day by day and month by month. They monitor natural and artificial environments all the time, and their observations are included in what will become the observed climatological record. Frequently, scientists will take field measurements for their research because that kind of information had not been gathered in a reliable or systematic way in the past.

Ships take measurements in the ocean while passing from one port to another. These "ships of opportunity" provide the monitoring and research communities with valuable information on locations where no reporting stations are permanently based. The oceans provide good examples of how difficult it is to gather information at all points in the world. To do so is quite expensive. The same is true for monitoring the climate system from the land. Data-recording equipment often breaks down in harsh environments and remains unrepaired for long periods. Because there is no way to protect them, these remote sites sometimes are vandalized and put out of action. When they are no longer operating, information is lost and the continuous record is broken.

Satellites now provide climate information on an hourly and daily basis. Their coverage is global, unlike observations taken from the ground. They do require a check on the ground (called ground-truthing) to make sure that what they report they are seeing is actually what they are seeing. Are they really monitoring vegetation or is it livestock? Is smoke disrupting estimates of the extent of forest fires? The advantages of using satellites are many, but there are also drawbacks: Satellite systems are expensive to maintain; they cannot monitor effectively under all atmospheric conditions; they require ground-truthing; and they produce more information than can be analyzed.

Scientists are often brought together to collect data, test hypotheses, and share their experiences in large-scale international climate experiments, some of which are coordinated by the World Climate Research Program in Geneva, Switzerland. The list of programs carried out over the years reads like an alphabet soup: GARP, GATE, EPOCH, CUEA, TOGA, COARE, PAGES, GOALS, CLIVAR, and so on. To better monitor and coordinate information, the international community is organizing a comprehensive global-observing system that focuses on climate, the ocean, and the land surface. This recently established overarching system is designed to monitor the key factors affecting global climate. Such observational activities are used to glimpse early changes in the climate system and in the environment.

The purpose of monitoring is to identify potential problem areas as early as possible, thereby providing enough lead time, it is hoped, to do something about them. Although some extreme climate events are closely monitored from space, aircraft, and the ground, we still are not certain about all of the factors involved in their creation, growth, and decay.

MONITORING / PALEOCLIMATIC CHANGES

"To know the road ahead, ask those coming back." This Chinese proverb applies to the study of climate change. In their attempts to understand past climate changes, scientists use a variety of research tools to reconstruct those climates. Indirect evidence of changes that occurred centuries and millennia ago is referred to as "proxy" information.

Some researchers rely on the evaluation and analyses of ice cores taken from the polar regions and glaciers to help them identify past climates. Dating of ice layers provides information on the deposition of certain types of marine and terrestrial life or volcanic aerosols and dust.

By analyzing the chemical composition of gases trapped in bubbles in the glacier ice, researchers can determine atmospheric levels of greenhouse gases at the time of entrapment. The Vostok ice core from the Antarctic spans a period of 160,000 years and provides an opportunity to identify correlations between carbon dioxide concentrations and temperatures of the atmosphere. Evidence shows that past variations in climate correlated well with levels of atmospheric greenhouse gases. The ratios of oxygen isotopes in the ice layers also provide information about prevailing temperatures at the times the layers formed.

Some researchers rely on studies of pollen and marine and lake sediments. Others, like Gordon Jacoby of the Tree-Ring Laboratory at Columbia University's Lamont-Doherty Earth Observatory, study tree rings. In the U.S. Department of Energy's 1986 *Proceedings of the International Symposium on the Ecological Aspects of Tree-Ring Analysis*, Jacoby suggests that "trees are almost unique in having absolutely datable

PETER MENZEL

Such records as fossilized fish (above) and tree rings tell us much about prehistoric climates, often providing insight into climate changes. The search for clues about past climates takes scientists to great heights and depths. To gather ice samples, a British scientist descends into a crevasse on Antarctica's Brunt Ice Shelf (opposite). The walls of such crevasses supply climatological data similar to that taken from ice cores.

annual growth increments that can respond to and record environmental changes." Variations in the width of tree rings correlate with changes in environmental conditions, and climatic variations account for a considerable portion of those changes. (Wide rings suggest wet growing seasons, and narrow rings indicate dry growing seasons.) In the world's dry areas, lack of precipitation limits tree growth, while in the colder and high-elevation areas, temperature becomes the limiting factor.

Scientists called palynologists have taken samples from lake sediments to identify the kinds of vegetation that existed at different periods in the planet's history. Their research has shown that 4,000 to 8,000 years ago the atmosphere was warmer by about 1.8°F. Pollen

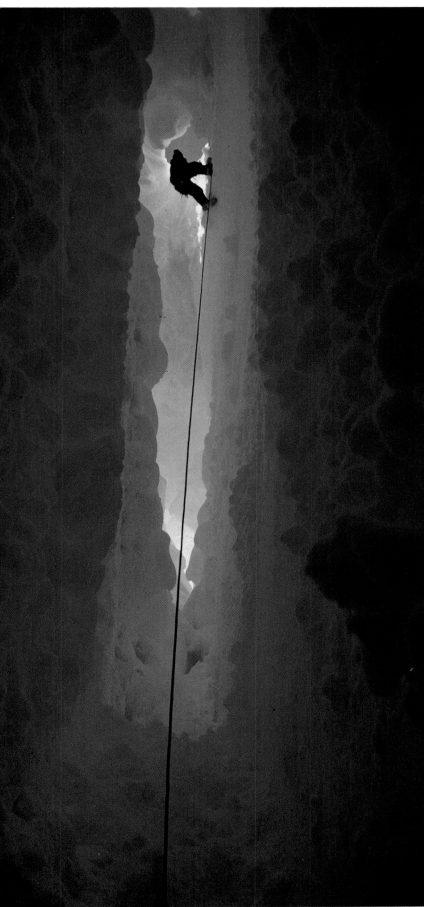

analyses of lake sediments suggest that the Great Plains, which are today the agricultural heartland of America, were much more arid and populated with plants that were more able to cope over long periods of low rainfall amounts. For their part, geologists also contribute to the reconstruction of climate history by looking for evidence in landforms and features. Floods, for example, leave their mark on the earth's surface, as do volcanic eruptions. Vegetation is often buried under mud or dust during rainy or arid periods. The effects of ancient flooding, such as mudslides, have been used to identify the frequency and intensity of prehistoric El Niño events in Peru.

Insight into past ocean temperatures is gained by analyzing marine sediment cores, which contain remnants of plankton and other organisms sensitive to small temperature changes. These organisms exist in large numbers; and when they die, they settle to the ocean floor, where they eventually form deposits in various layers of the seabed.

The late J. Murray Mitchell, a well-respected climatologist who worked for many years at the National Oceanic and Atmospheric Administration (NOAA), put great stock in paleoclimatic research. In his foreword to A. D. Hecht's 1985 book, *Paleoclimatic Analysis and Modeling,* Dr. Mitchell advised that "the fruits of paleoclimatic studies are many and by no means lacking in relevance to practical, modern-day concerns about climate. They help us to understand the capacity of climate to vary the rates of change and the degree of consistency and regularity. They also illuminate the causes of climatic change."

MONITORING / ANALOGUES

Historical records, diaries, travelogues, and paintings can provide insights into climate history. This material can be viewed as proxy information, but it is also considered anecdotal and less reliable than today's precise quantitative measurements.

Historical records have been used just about everywhere to gain a glimpse of past climates. Emmanuel Ladurie, for example, in his book that discusses climate since the year 1000, drew on the harvest dates of grapes—recorded by European monasteries—and wheat. In earlier centuries of this millennium, wine and bread were two extremely important products.

During the Cultural Revolution in the 1960s, China's universities were closed. Nevertheless, Chinese professor Wang Shao-wu analyzed many of his country's historical records to create detailed, year-by-year maps of droughts and floods for the past 500 years.

Studies of Canada's climate focused on personal diaries and shipping records that noted ice conditions in Hudson Bay. The timing and degree of ice cover served as an indirect indicator of conditions over decades and centuries.

Similar information has been used to reconstruct El Niño events, which are known to be associated with droughts in Indonesia. Records from cash-crop plantations were closely analyzed to learn when droughts reduced production, thereby suggesting the possibility of El Niño. In the early 1890s, Peruvian Victor Eiguguren collected information on the frequency of heavy rainfall in normally arid areas along Peru's coast, knowing that heavy rains there occurred only during such events. Travelers' journals, church records, and accounts of "abnormal" weather were used to reconstruct the history of El Niño.

Around the same time, American geologist Alfred Sears studied vegetation along Peru's coastline. He provided insight into the activities of Spanish conquistador Francisco Pizarro, basing the following account on historical records:

"It rains on the northern coast of Peru only once in seven years. All the remaining years are utterly dry. Pizarro could not have gone from the Tumbez, where he first landed after his fight with the Indians at Puna, to the valley of Tangarara and found feed for his animals, nor would he have found the little settlements mentioned as existing along the road in any other season than a wet one. Again, he would not have been driven from the Chira Valley, Tangarara, by malarial fevers, save in a wet year, as it exists in the valley only in wet years."

Sears's report suggests that Pizarro came ashore during El Niño. Had he done otherwise, his landing likely would have failed and Peru's history might have been entirely different.

Other proxy information comes from archaeological sites where anthropologists gather material about foods that were eaten, the types of agricultural systems that were used, and the water resources that were available.

Accounts in poems, proverbs, prose, and art also provide insights into past climates. For example, a book on the history of meteorology provides the following example from the Egyptian "Hymn to the Sun" by Pharaoh Akhnaten (1370 B.C.): "and you bring life to distant lands, giving them a Nile in the sky, which falls to the earth and drenches the hills with its waves, like an enormous green sea, and waters

Scientists at Nevada's Desert Research Institute sample wine from a vineyard in California's Napa Valley. They look for differences in ratios of light and heavy isotopes of hydrogen and oxygen; rain, wind, and other climate-related factors cause ratios to vary. The information provides more detailed data than normal weather records do, helping to validate computer models that predict temperature and precipitation.

STEVEN PUMPHREY

the fields." The hymnal reference is to Syria, which at that time was under Egyptian rule and considerably wetter than arid Egypt.

Historical Analogues

Social scientists have begun to identify how society might prepare for climate fluctuations over decades or for climate change over even longer periods. Some researchers rely on modeling activities that produce scenarios of what might happen with different degrees of global warming (based on projected levels of greenhouse gas emissions). Although present models provide only crude pictures of the regional impacts of global warming, researchers believe they can learn much about how great the impacts could be on activities such as farming, ranching, commerce, and settlement.

How can we protect society? We can draw on our historical experiences and the experiences of others in coping with similar climatic events. We can glimpse the future, for example, by looking back on how well we coped with Hurricane Hugo or the Great Midwest Flood. We can recall how land-use practices in the early 1900s contributed to the Dust Bowl of the 1930s. We can review the environmental impacts in the 1980s of deforestation in Brazil's state of Rondônia. History is rich with lessons.

Social scientists have sought to work from the societal end of the chain of extreme events rather than from the physical end. They have identified historical analogues to what a global warming might be like. They argue that if we are not coping well with variability under today's global climate regime, how can we be expected to perform better in future decades when faced by an altered climate regime.

Physical scientists say that drought-prone areas likely will become more so in a warmer climate and that wet areas will get wetter. Social scientists can look back at recent decades to identify severe droughts or floods to uncover society's potential weaknesses and strengths in coping with possible global warming impacts. These situations, of course, only suggest the possible impacts of climate change and how society might respond to them. But they provide information on how well we are coping with today's extreme events. This is a "win-win" situation: We can use the information to enhance society's ability to cope more favorably with an uncertain future, whether or not the global climate changes.

Whether the climate gets warmer or cooler or stays as it is today, there will be variability. Assessments of how well societies cope with variability can only serve to add to our body of knowledge about climate-society interactions.

CLIMATE SURPRISES

Clouds of steam and other gases rise from the Lenin Steel Mill in Poland. As fossil-fuel burning and land-use activities such as tropical deforestation release greenhouse gases and thereby heat the atmosphere, unexpected twists and turns in the climate system may result. Scientists who try to anticipate what the future holds say they expect some nasty surprises—and perhaps even a few favorable ones.

Each place has a climate-related hazard to worry about: flooding, drought, high winds, fires, tornadoes, hurricanes, tidal waves, blizzards, ice storms, hail, and so on. While residents may be generally familiar with their local hazard, they are not sure when it will occur or how intense it will be. As a result, people are often surprised when it becomes reality.

Surprise can be defined as a gap between expectation and what happens. In a sense, there are expectable surprises as well as unexpectable

ones. For example, people who live in an area susceptible to flooding know it, and if someone were to tell them a flood was possible, they would acknowledge that as a fact. No surprise here. What they do not know is when flooding will take place or how bad it will be. These occurrences are knowable and expectable, but their intensity and timing may be surprising.

Unexpectable surprises are the most worrisome. The thinning of the stratosphere's ozone layer, popularly referred to as the ozone hole, is a good example of an unexpectable surprise. Scientists had no way to predict it.

Unexpected twists and turns in regional climate might result from fluctuations or global warming. One surprise was the set of hard freezes in Florida's citrus region during the hottest decade on record. Another surprise might be that while summers get hotter, winters could get colder—not warmer, as might be expected.

Surprises represent a dread factor in the issue of climate change. Getting policymakers to change the status quo regarding industrial or agricultural activities that produce greenhouse gases is a difficult task, given the uncertainty about global warming and its possible impacts. Over the years, several surprises that might result from global warming have been suggested, and scientists who believe that human activities are already altering the climate are concerned about surprises like the ozone hole.

Surprises include the possible breakup of the West Antarctic ice sheet, a drastic change in ocean currents, rapid changes in sea level, summerless years, and counterintuitive climate anomalies like untimely freezes in Florida.

How fast the world's climate changes is very important, and national policymakers are increasingly warned about the possibility of abrupt changes from global warming. Abrupt and unusual changes in climate are within the realm of possibility.

To nonscientists, the increase in greenhouse gases appears to be small—on the order of one part per million by volume of CO_2 each year, or less than half a percent per year. This number is so small that it can lead us to think that climate changes resulting from such increases also will be small. But the geological record shows that past climates did not just change slowly over thousands of years. Once a threshold is reached, abrupt change can occur. While a doubling of CO_2 over pre-1750 levels has been chosen by researchers as a yardstick to identify rates of change, it is not a significant threshold after which some spectacular environmental change will occur. Scientists do not yet know the threshold of concentrations that must not be exceeded.

One cause of surprise is not so much the event itself but our knowledge about it. Scientists have been collecting information for a relatively short time. In parts of West Africa's Sahel region, for example, the record of reliable information about rainfall may be only five decades long. In the United States, the record for the Great Salt Lake is about 150 years. For some European cities, the record could be 250 to 300 years. Our reliable data on climate variability and change are very short, so we should not believe we have experienced all the possibilities in a region. Nevertheless, humans tend to think that climate situations of the recent past are likely to last into the future. This tendency causes people to be surprised when an unusual change occurs in the conditions we have come to expect.

When scientists discover new information about how the climate system works, a piece is added to the puzzle. The more they learn, the less often people will be surprised. Because we do not have complete information on how the system and its components interact, we must accept the possibility of surprises. While we do not know when surprises might occur, we can learn from past climates about the kinds of surprises we could be subjected to in the future.

SURPRISES / OZONE HOLE

Scientists have worried about thinning of the ozone layer (some say ozone shield) since at least the early 1970s, because depletion of stratospheric ozone would allow more biologically destructive ultraviolet radiation (UV-B) to reach the earth's surface. Medical experts have noted that an increase in UV-B exposure would result in more cases of skin cancer and cataracts in humans.

In the mid-1970s atmospheric chemists F. Sherwood Rowland and Mario J. Molina at the University of California at Irvine realized that chlorofluorocarbons (CFCs) could enter into a catalytic chain reaction with ozone in the upper stratosphere. Doing so would damage the ozone layer, thereby reducing protection for living things on the earth's surface. Following the discovery, V. Ramanathan, a researcher with the National Aeronautics and Space Administration, found that CFCs were powerful greenhouse gases. Later it was learned that CFCs contributed about 20 percent to the warming potential generated by human activity. At that time, however, the prospect of enhancing the naturally occurring greenhouse effect was not of much concern. Interest in the CFCs really centered on the possibility that human activities could lead to a depletion of stratospheric ozone.

Environmental concerns about CFCs were heightened because of their use in products such as hair spray and deodorants. They also were used as refrigerants, foam-blowing agents, and cleansing agents for computer chips. In the late 1970s, the use of CFCs in spray cans was voluntarily discontinued by several countries, including the United States. While CFCs are

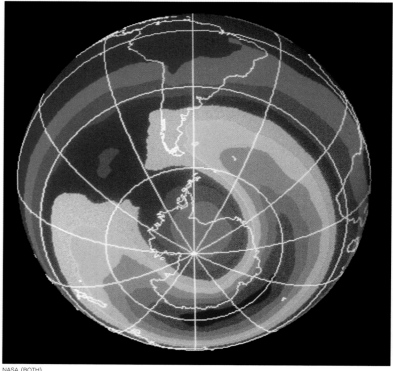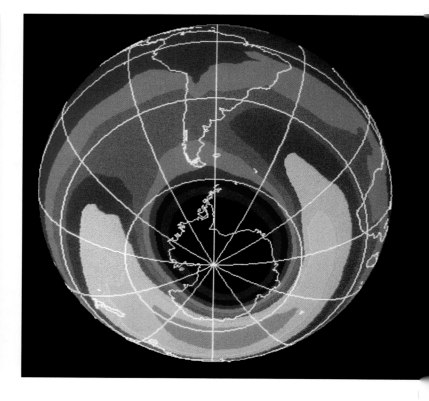

NASA (BOTH)

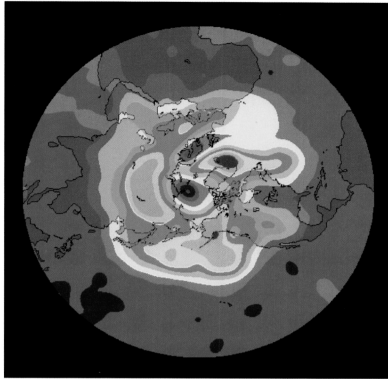

Widespread use of chlorofluorocarbons (CFCs) thinned the atmosphere's ozone layer. Color-enhanced satellite images for 1979 (opposite, left) and 1989 (opposite, right) highlight in black the change in ozone concentration above Antarctica. A 1995 image (above), shows thinning over a smaller area (bull's-eye position) of the Northern Hemisphere.

inert in the lower atmosphere, within a few years they can rise into the stratosphere, where they can survive for a couple of centuries. The CFC molecules decompose, however, when they are struck by ultraviolet rays, releasing the chlorine atom that attacks ozone in catalytic chain-reaction fashion. This atom can survive and attack thousands more ozone molecules.

Several governments, however, were not yet convinced that CFCs posed a danger or that a possible problem outweighed their utility to society. As a result, production rose again in the early 1980s, following a sharp drop in the late 1970s when some countries decided to discontinue their use as spray-can propellants.

Government representatives came together in Vienna, Austria, to draw up an agreement for controlling CFC production and use. Attitudes changed radically with the 1985 discovery of the infamous ozone hole over the Antarctic.

In 1985, Joseph Farman and colleagues from the British Antarctic Survey published an article on stratospheric ozone measurements. They had discovered that October ozone concentrations over the Antarctic region had dropped by about 40 percent since the 1960s. Research findings strongly linking CFCs to ozone destruction in the polar regions startled the scientific community. As reported in a National Academy publication in the 1990s, "most scientists greeted the news with disbelief. Existing theory had not predicted it." In fact, very low ozone values had been recorded by a satellite instrument during the Southern Hemisphere's springtime, but American and British scientists rejected the readings until the early 1980s, considering them the result of instrument malfunction.

This "accidental" discovery surprised policymakers, prompting them to put together an international agreement in 1987—the Montreal Protocol on Substances that Deplete the Ozone Layer—to restrict industrial production and use of CFCs. Later amendments called for a phaseout of CFC usage. Banning CFCs would yield a double benefit: protect stratospheric ozone and reduce their contribution to climate change.

Because they can survive for decades in the atmosphere, it will be at least a few decades before CFCs will be reduced enough to begin the process of restoring ozone to its earlier normal levels. In the meantime, CFCs will continue to destroy stratospheric ozone. In the first half of the 1990s, ozone levels over the Antarctic set new minimum low levels with a 60 percent loss compared with pre-1980 levels.

SURPRISES /CURRENT FLIP-FLOPS

According to Columbia University professor Wallace Broecker, a major ocean current moves around the globe like a conveyor belt. He describes the process associated with the "belt" as follows: A shallow, warm current flows toward the North Atlantic from the Pacific Ocean. Once it gets there, Broecker says, the water cools and thus becomes relatively more dense. The cool, dense water begins to sink, eventually becoming part of a deep ocean current flowing southward toward the Pacific.

In the Pacific region, the current of water begins to warm and rises to the surface where it warms some more and moves along again in a shallow northward-flowing current. According to Broecker, "the water that sinks to the bottom of the northern Atlantic flows down the full length of the Atlantic, around Africa, through the southern Indian Ocean, and finally up the Pacific Ocean."

The implications of the movement of warm surface water to the North Atlantic region are clear. It heats up the atmosphere above the ocean as well as the areas near its shores, giving Paris, for example, the same average temperatures as New York City, although Paris is several degrees latitude farther north. The conveyor carries a considerable amount of heat to the North Atlantic region. As water evaporates and cools in the Atlantic, the seawater becomes more salty and dense, thereby driving the Pacific-ward flow of the oceanic conveyor belt.

This is the situation as we currently understand it. Scientists think that if global warming were to take place the surface waters of the North Atlantic might remain warm instead of cooling and sinking. The conveyor belt would be shut off. Broecker suggests "it is possible that surface warming and a permanent melting of sea ice would suppress the sinking and insulate the deep ocean water from the upper layers of the ocean and the air above it."

This result could be catastrophic for marine life, because it would drastically change the ocean currents and regional temperatures. It would also reduce the ability of the world ocean, as a buffer to climate change, to absorb heat from the atmosphere.

The ocean serves as a sink for carbon dioxide (CO_2); that is, it pulls CO_2 out of the atmosphere and then buries it for several centuries in the deep ocean. If the conveyor belt were shut off, the ability of the ocean to absorb and sequester CO_2 for a millennium or more would decrease, and global warming would increase. On land, regional temperatures and climate anomalies would also change.

While all this scientific discussion of oceanic processes may seem far-fetched, geological and other proxy (indirect) evidence, such as the historical climate information obtained from ice cores, suggests that the conveyor belt has turned off in the past for one reason or another. Proxy information also suggests that switches in the conveyor belt's operation were accompanied by sharp increases in carbon dioxide in the atmosphere. Such catastrophic flip-flops could take place not in centuries or millennia, but in a matter of only a few decades. The oceanic conveyor belt is being closely scrutinized by climate researchers, in light of the serious worldwide repercussions if such a flip-flop were to occur abruptly in the 21st century.

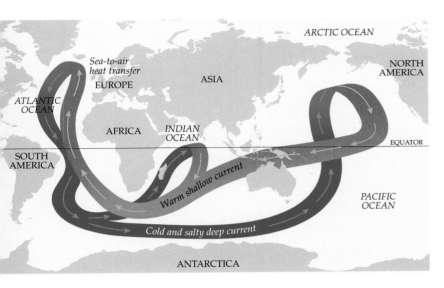

Currents move through oceans like a conveyor belt (left). A sudden flip-flop or shutdown could severely alter climates tempered by warm currents. Changes in ocean currents could disrupt naturally occurring CO_2 sinks, such as the ones shown by the white streaks in this photograph of the Little Bahama Bank (below).

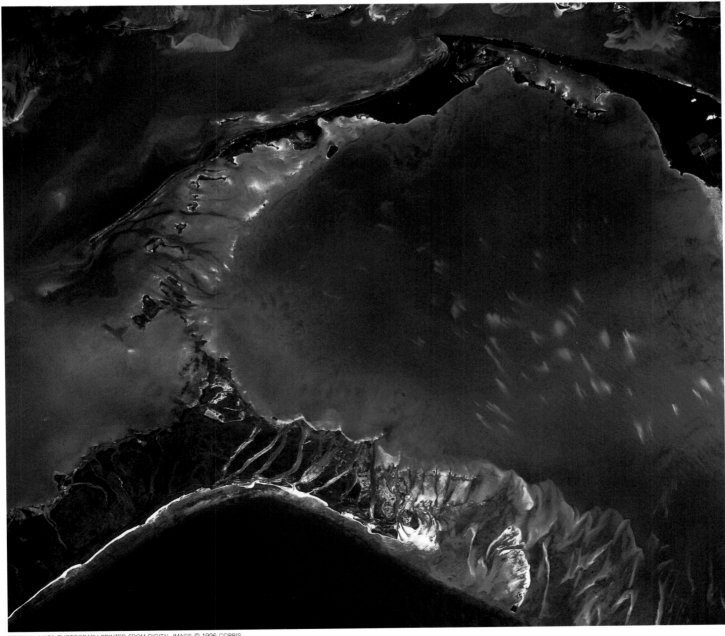

SURPRISES / FLORIDA FREEZES

The state of Florida is well known for its many orange groves. The mid-section of the Florida peninsula was once covered with highly productive orange groves owned and run by families and large food conglomerates. In California, oranges are grown mainly to be eaten as fruit, but in Florida, oranges are generally grown to make one of the most popular breakfast drinks in the United States. John McPhee, in 1966, captured the interesting life of the Florida citrus growers in his book, *Oranges*.

The oranges from Florida are converted into frozen concentrated orange juice (FCOJ), which commands a high price as a commodity in the world marketplace. Prior to a set of devastating freezes in the 1980s, central Florida was considered an ideal location for growing citrus, whereas the state's southern half was not so favorable to growers. The soils in the south are heavy and wet and, therefore, not as healthy for the roots of orange trees. As a result, the citrus industry came to be concentrated in the central part of the state.

The weather in central Florida is not always perfect, however. Citrus growers know that Florida is unfortunately prone to frosts and hard freezes. While an untimely frost may damage the fruit while it is still hanging from the trees, a hard freeze (very cold temperatures) even for a short time can actually kill the orange trees. Growers can be hurt by frosts in the sense that they might lose part or all of their crops after just a few hours of frost or chilling temperatures. Hard freezes, however, can destroy the trees bearing the fruit and bring growers to economic ruin. Even if growers are financially able to replant orange trees in their orchards, the young trees usually take about four or five years before they begin to bear fruit.

A couple of decades ago, the growers had a few options that they could pursue to combat nature's negative impacts. For one thing, many of them owned more than one grove so that in the event of a frost, the oranges in at least one of the groves might survive. If a frost damaged a crop, the prices for the remaining oranges would increase sharply and a grower could still survive economically.

During one memorable freeze, I watched television coverage of worried growers using smudge pots and burning rubber tires to ward off chilling temperatures that would have damaged their orange crops or trees. Spraying trees with water is another tactic growers use in their attempts to protect their groves. As the water cools, it releases latent heat that keeps both the fruit and the trees warmer than if they were dry.

In 1963, there was a major freeze in the Florida citrus area, and the price for orange juice shot up. This business opportunity immediately caught the attention of Brazilian investors, who jumped at the chance to produce FCOJ and ship it for sale in the United States. The availability of oranges from Brazil meant that the prices in the future would not fluctuate so much, and that losses of oranges by the American growers could not eventually be made up by higher prices at the market. Today, Brazil is one of the major exporters of FCOJ to the United States, Europe, and elsewhere. This economic change has made Florida's citrus growers even more vulnerable to the vagaries of the weather.

Record-setting freezes in the 1980s surprised Florida citrus growers, often wrapping their orange groves in wintry coats of ice. These same years, however, made history as the world's hottest decade on record. While the average global temperature may increase, regional and local temperatures may be warmer—or colder.

In the 1980s, Florida was unfortunately plagued by several hard freezes that destroyed many of its orange groves. In fact, the number of Florida freezes during those years broke a 130-year-old record for the highest number of freezes in one ten-year period.

Despite what happened, in the mid-1980s the U.S. Environmental Protection Agency (EPA) decided not to give its support to a proposed study of the impact of a Florida freeze on oranges, citrus growers, and consumers as an example of the kind of climate-related surprise people might come to expect if the global climate were to warm up over the next several decades. The agency declined its support because its officials believed, perhaps somewhat logically, that a global warming would effectively get rid of the threat of additional freezes. However, the stress of the severe freezes (during what turned out to be the hottest recorded decade) showed that counterintuitive surprises are likely during global warming. In such a warming situation, average temperatures will certainly increase, but the number of extreme weather events might also increase both in summer and in winter.

The experience of frequent Florida freezes in the 1980s strongly suggests that there are potential surprises in the global climate system as it undergoes change. These surprises may come not only from global warming (climate change), but they may also come from aspects of climate variability that we have not yet experienced or even anticipated. The historical record is not long enough for scientists to have a clearer understanding of what might happen in the future.

SURPRISES / HOLLAND FLOODS

Growing up in Rhode Island in the early 1950s, I often went to the movies and watched newsreels on world events. I specifically remember seeing a film about devastating floods in Holland (the Netherlands). In 1953, severe storms took place along the coast of that country, and the dikes that had held back the North Sea for centuries were breached. The catastrophic flooding that followed claimed more than 1,800 victims and destroyed almost 50,000 homes.

The floods that had surprised the Netherlands were unprecedented, as a Dutch government report noted, "not only because they were so sudden but also because they brought with them so much water in such terrific force." The dike system had worked well for such a long time that no flood of comparable size had occurred since 1570. People living near the coast and on islands just offshore had grown secure behind the dikes that had protected them and generations of their families. Now and then, some water had breached the dikes, but pumping systems had always dealt with it quickly.

The conditions for flooding began on the evening of January 31, 1953. At dusk, survivors remembered later, they could hear the winds howling and the sea roaring with rage. More than a score of ships along the coast signaled distress as they rode out the storm, but there was little anyone could do to help them. When the tempest subsided a day later, the devastation it left behind was immense, and its toll was staggering. A year would pass before some refugees from the flood could return home.

Today, about 40 percent of Holland is below sea level, with an estimated 1.5 million acres having been reclaimed from the sea. The government spends hundreds of millions of dollars a year to keep the country dry, often reminding us that "without the man-made earthen dikes and pumping stations, over half the country would disappear overnight."

After the 1953 floods, engineering skills were called on again. The dikes were rebuilt bigger and better than they had been before as part of the government's 30-year, multibillion-dollar plan to protect the country from North Sea storm surges. The last phase of that plan was the completion of a movable barrier to protect the southwestern part of the country. In the spring of 1990, the system was tested by the sea, but the barrier withstood the force of nature. "The surge barrier passed with flying colors and should protect the Netherlands for centuries to come," one observer noted.

In recent years, the possibility of global warming and rising sea levels has concerned Dutch scientists, who fear that the catastrophic flooding they had previously expected to occur once in 10,000 years could instead happen every 1,000 years.

The Netherlands again suffered severe flooding in 1993—not from the sea, but from one of its two major rivers, which overflowed its banks and inundated much of the country. The Meuse had been at floodstage for two years when its levees finally gave way. Fortunately, the Rhine's dikes held, but the threat of flooding prompted the biggest evacuation in its history: A quarter of a million people fled their homes. High water in the rivers resulted from heavy rainfall—and containment of the rivers by engineering construction, or so they had thought. While the Dutch kept their eyes on levees holding back the ocean, they were blindsided by river flooding, a natural hazard that they thought was under control. One news report noted, "We Dutch have been so busy looking after our front door to the sea that we forgot about the back door to the big rivers."

A costly program to strengthen river dikes, which Parliament had held up for years, was quickly approved. It struck a balance between building broader and higher dikes, while preserving old homes and landscapes behind them.

Weary residents of the Netherlands, beset by devastating floods in 1993, endure an age-old struggle against high water. This time, however, the threat came not from the sea, held back by extensive dikes. Torrents of water overflowed the Meuse, one of the country's two great rivers. Swollen with rain, it broke through levees and swept across the land.

In the United States, Utah's Great Salt Lake is the largest lake west of the Mississippi River. This well-known body of water covers an area about the size of the state of Delaware. As a terminal lake, all of the water that flows into it from snowmelt in the mountains has no natural means of escape except through the process of evaporation in the summertime. When evaporation is greater than precipitation and runoff, the lake drops.

During more than 150 years of record-keeping, the level of the lake has fluctuated within a range of less than 20 feet. The decades of the 1950s and '60s were periods of low lake levels, and in 1963, the Great Salt Lake dropped to its lowest level since 1840. As the level declined, many people (including scientists) thought it would never return to the higher levels of earlier decades.

Thinking that the decline was a permanent one because the level of the lake had been fairly stable for a few decades, developers requested and got permission to build on the newly exposed lake bed. Writer Rick Gore succinctly described the onset of the problem in the June 1985 issue of National Geographic: "Record rainfall in September 1982 triggered the basin's ordeal. The winter brought unprecedented snows, as great as 835 inches, to the Wasatch. A sudden May heat wave sent torrents of snowmelt down the slopes toward the lake."

From 1982 to 1986, the lake level increased by 12 feet as a result of snowmelt from the unusually heavy snowfalls that had piled up in the Wasatch Mountains. The water flowing into the lake was more than 200 percent of normal for each of the years between 1983 and 1986.

The sharp increase during these years was totally unexpected by scientists, policymakers, and the public. Their responses each year were based on their belief that the heavy snows were unusual. Salt Lake meteorologist Mark Eubank remarked at the time that "we thought the first winter was a once-in-a-hundred-year-event. Then nature turned around and did the same thing again. You'd never predict that, never expect it."

Almost everyone was certain that the lake level in the following year would return to normal. So, the track beds for the railroads were raised a few feet each year to accommodate the rising water level. And each year, as the lake's level steadily increased, more buildings, roads, and railway tracks went underwater. Each year people thought they would see the last of the increase in lake level.

To prevent further destruction in and around Salt Lake City, officials of the state government decided to build a pumping station that would pump the excess water into surrounding dry depressions and away from the city. The year after the station was completed, however, the level of the lake began to drop and the threat disappeared.

This situation provides interesting insights into some of our perceptions about the natural environment. The climate is variable on different time scales, from years to decades. The levels of the lake also fluctuate in response to the variations in climate. Not expecting such great variations in a short span of years, the people of Utah were surprised when the Great Salt Lake dropped to its lowest level in the early 1960s, and they were equally surprised when the lake

High water closed beaches along Utah's Great Salt Lake in the early 1980s. The level of the lake astonished everyone by climbing an unprecedented 12 feet higher from 1982 to 1986, inundating railroad tracks, roadways, and buildings near the shore—such as the century-old Saltair resort (background), where rising waters crept in and occupied many of the rooms.

JIM RICHARDSON

levels rose sharply in the early 1980s. The late historian Dale Morgan seems to have comprehended the lake best when he said, "It is intolerant of men and reluctant in submission to their uses."

Today, planners are aware that the lake will naturally fluctuate in size and that they must take that into account in their urban planning activities. Although there is plenty of uncertainty in forecasting future lake levels, decision-makers can reduce the element of surprise with proper planning around the fringes of the Great Salt Lake, and with proper respect for normal lake-level variations.

NOTES ON CONTRIBUTORS

H. J. de Blij, a member of the National Geographic Society's Committee for Research and Exploration, appears frequently on NBC News as a geography analyst. He is University Scientist at the University of South Florida in St. Petersburg and the author of some 30 books, including *Physical Geography of the Global Environment.*

Michael H. Glantz is a senior scientist at the National Center for Atmospheric Research, in Boulder, Colorado. For his work in the field of climate impacts, he received the United Nations Environment Programme's Global 500 award. He has edited or written 17 books, the latest of which is *Currents of Change: El Niño's Impact on Climate and Society.*

Stephen L. Harris, chair of the Department of Humanities and Religious Studies at California State University, Sacramento, has written about geologic hazards in *Agents of Chaos* and *Fire Mountains of the West.* Other publications include *Understanding the Bible* and *Classical Mythology: Images and Insights.*

Patrick Hughes, a freelance writer, was managing editor of *Weatherwise* magazine from 1986 to 1991. He has also been a meteorologist and writer-editor for the National Oceanic and Atmospheric Administration. His publications include some 90 articles and 2 books on meteorology and weather-related subjects. In 1995, he coauthored *Meteorology and America, 1920-1995,* an enlarged, book-format edition of *Weatherwise.*

Richard Lipkin, the 1997 recipient of the American Chemical Society's James T. Grady–James H. Stack award, is a freelance science writer based in New York City. A coauthor of *Nature on the Rampage* and *Mathematical Impressions,* he has served as writer and editor for *Science News, The Wilson Quarterly,* United Press International, and The MacNeil-Lehrer Newshour.

Jeff Rosenfeld has been an editor of *Weatherwise* magazine for more than a decade. A science journalist with a background in history, he has written extensively about the latest atmospheric research as well as infamous storms of the past. He has received several awards from the Educational Press Association of America.

Richard S. Williams, Jr., is a research geologist with the U.S. Geological Survey in Woods Hole, Massachusetts, and vice chairman of the National Geographic Society's Committee for Research and Exploration. He is a specialist on satellite remote sensing of dynamic geologic, glaciologic, and geomorphic processes and the author of more than 150 papers, book chapters, and maps. He serves as senior editor and contributing author of the 11-volume series, *Satellite Image Atlas of Glaciers of the World.* Antarctica's Williams Glacier is named for him.

ACKNOWLEDGMENTS

The Book Division wishes to thank the many individuals quoted or mentioned in this publication for their help and guidance. In addition we are grateful for the assistance of the following people: Dan Albritton, Neil W. Averitt, Johan Bouma, Mark Cane, Martha C. Christian, Jerry Eaton, Stanley Gedzelman, Thomas P. Grazulis, Patrick Hughes, Dale Jamieson, Arch Johnston, Marianne R. Koszorus, William Mass, Kathleen Miller, Jack Morris, Mikiyasu Nakayama, Neville Nicholls, Roger Pielke, Tom Potter, Thomas W. Schlatter, Jagadish Shukla, Xiaodong Song, D. Jan Stewart, Will D. Swearingen, Jim Titus, Joe Tribbia, Sergei Vinogradov, J. Michael Wallace, Don Wilhite, Richard S. Williams, Jr., and George Woodwell.

ADDITIONAL READING

The reader may wish to consult the *National Geographic Index* for related articles and books. The following titles may also be of interest:
Bruce A. Bolt, *Earthquakes* and *Earthquakes and Geological Discovery;* Raymond S. Bradley and Philip D. Jones, *Climate Since A.D. 1500;* Fred Bullard, *Volcanoes of the Earth;* Robert W. and Barbara B. Decker, *Mountains of Fire* and *Volcanoes;* Myron L. Fuller, *The New Madrid Earthquake;* Thomas P. Grazulis, *Significant Tornadoes 1680-1991;* Gladys Hansen and Emmet Condon, *Denial of Disaster;* Stephen L. Harris, *Agents of Chaos;* Robert L. Kovac, *Earth's Fury;* H. H. Lamb, *Climate, History and the Modern World;* David Lambert, *The Field Guide to Geology;* Charles C. Plummer and David McGeary, *Physical Geology;* Frank Press and Raymond Siever, *Earth;* Andrew Robinson, *Earth Shock;* David Roland, ed., *Nature on the Rampage;* Keith Smith, *Environmental Hazards—Assessing Risk and Reducing Disasters;* Time-Life Books, *Flood* and *Storm;* Kaari Ward, ed., *Great Disasters;* Jack Williams, *The Weather Book;* and Chen Yong et al., *The Great Tangshan Earthquake of 1976.*

Library of Congress CIP Data
Restless earth / prepared by the Book Division, National Geographic
   Society, Washington, D.C. ; [foreword by H.J. de Blij].
    p.   cm.
   Includes bibliographical references and index.
   ISBN 0-7922-7026-6. —ISBN 0-7922-7006-1 (deluxe). —ISBN
0-7922-7121-1 (trade).
   1. Climatology. 2. United States—Climate. 3. Earth.
4. Volcanoes.   I. National Geographic Society (U.S.). Book
Division.
QC981.R47 1997
551.6—dc21   97-8115

Composition for this book by the National Geographic Society Book Division. Printed and bound by R. R. Donnelley & Sons, Willard, Ohio. Color separations by World Color Graphics, Washington, D.C. Dust jacket printed by Miken Companies, Inc., Cheektowaga, N.Y.

Visit the Society's Web site at
www.nationalgeographic.com.

INDEX
Boldface indicates illustrations.